CAMBRIDGE TRACTS IN MATHEMATICS

General Editors

---

127 Spectral Theory of the Riemann Zeta-Function

CAMBRIDGE TRACTS IN MATHEMATICS

127

Yoichi Motohashi
Nihon University

Spectral Theory of the Riemann Zeta-Function

CAMBRIDGE UNIVERSITY PRESS
Cambridge, New York, Melbourne, Madrid, Cape Town, Singapore, São Paulo

Cambridge University Press
The Edinburgh Building, Cambridge CB2 8RU, UK

Published in the United States of America by Cambridge University Press, New York

www.cambridge.org
Information on this title: www.cambridge.org/9780521445207

First published 1997
This digitally printed version 2008

A catalogue record for this publication is available from the British Library

Library of Congress Cataloguing in Publication data

Motohashi, Y. (Yoichi)
Spectral theory of the Riemann zeta-function / Yoichi Motohashi.
p. cm. – (Cambridge tracts in mathematics; 127)
Includes bibliographical references
ISBN 0 521 44520 5 (hc)
1. Functions, Zeta. 2. Spectral theory (Mathematics)
I. Title. II. Series.
QA246.M78 1997
512′.73–dc20 96-46059 CIP

ISBN 978-0-521-44520-7 hardback
ISBN 978-0-521-05807-0 paperback

Contents

Preface

EVER since Riemann's use of the theta transformation formula in one of his proofs of the functional equation for the zeta-function, number-theorists have been fascinated by various interactions between the zeta-function and automorphic forms. These experiences, however, have remained episodic like rare glimpses of crests, for most of them ensued from apparently spontaneous relations of the zeta-function with a variety of Eisenstein series. Nevertheless such glimpses are highly suggestive of a grand view over and far beyond the Eisenstein ridge, and bring forth the notion of a *kamuy-mintar* where the entire collection of automorphic forms contribute to the formation of the zeta-function.

My aim in the present monograph is to try to substantiate this belief by demonstrating that the zeta-function has indeed a structure tightly supported by all automorphic forms. The story begins with an unabridged treatment of the spectral resolution of the non-Euclidean Laplacian, and continues to a theory of trace formulas. The fundamental means thus readied are subsequently mustered up for the quest to find an explicit formula for the fourth power moment of the zeta-values. Then the zeta-function emerges as a magnificent peak embracing infinitely many gems called automorphic L-functions representing the spectrum.

My best thanks are due to my friends A. Ivić and M. Jutila for their unfailing encouragement, and to D. Tranah, P. Jackson, and all of the personnel of the Cambridge University Press engaged in this project for sharing their professional vigor. I must also thank my family for the comfort and the music that have been sustaining my scientific life.

Tokyo
January, 1997 Y. M.

Convention and assumed background

Once introduced most symbols will remain effective throughout the sequel. Some of them are naturally standard. Thus $\mathbb{Z}$, $\mathbb{Q}$, $\mathbb{R}$, $\mathbb{C}$ are sets of all integers, rationals, reals, and complex numbers, respectively. For example the group composed of all $n \times n$ integral matrices with determinant equal to 1 is denoted by $SL(n, \mathbb{Z})$. The arithmetic functions $\sigma_a(n)$ and $d_k(n)$ stand, respectively, for the sum of the ath powers of divisors of n and for the number of ways of expressing n as a product of k integral factors. In particular, $d(n) = d_2(n)$ is the divisor function. The Bessel functions are denoted by I_ν, J_ν, K_ν as usual. We use the term K-Bessel function to indicate K_ν without the specification of the order ν; and the same convention applies to other Bessel functions as well. The symbol Γ is for the gamma function, and $\mathit{\Gamma}$ is for the full modular group introduced in Section 1.1. The dependency of implied constants on others will not always be explained, since it is more or less clear from the context.

Some knowledge of integrals involving basic transcendental functions is certainly helpful. For this purpose Lebedev's book [38] is quite handy. But there are occasions when Titchmarsh [69], Watson [74], and Whittaker and Watson [75] give more precise information, though proofs of most integral formulas and relevant estimates are given or at least briefly indicated either in the text or in the respective notes. In addition to these standard books, Sonine's article [68], Vilenkin's book [71], and the table [15] by Gradshteyn and Ryzhik are recommended.

Readers are supposed to have ample knowledge of the zeta-function such as that developed in Titchmarsh [70] as well as in Ivić [19]. In fact this monograph is, in part, a continuation of their books. Thus, for

instance, bounds like

$$(\log t)^{-1} \ll |\zeta(1+it)| \ll \log t, \qquad \zeta(\tfrac{1}{2}+it) \ll t^{\frac{1}{6}} \log t,$$

$$\int_0^T |\zeta(\tfrac{1}{2}+iu)|^4 du \ll T(\log T)^4 \quad (t,\ T \geq 2)$$

are used freely under the term *classical estimates.* On the other hand no experience in the theory of automorphic functions is assumed. The first three chapters can be taken for an introduction to the subject.

The references are limited to the essentials. Suggestions for further readings may be found in the notes and the articles quoted there.

1
Non-Euclidean harmonics

Our story of the Riemann zeta-function is to be unfolded on a stage filled with non-Euclidean harmonics. Accordingly we need first to tune our principal instrument. We are going to prove in this initial chapter a spectral resolution of the non-Euclidean Laplacian

$$\Delta = -y^2\big((\partial/\partial x)^2 + (\partial/\partial y)^2\big)$$

with minimum prerequisites. The entire theory originated in a seminal work of H. Maass, which was later developed by W. Roelcke, A. Selberg, and many others. Our account is an elementary approach to their theory in the case of the full modular group. Despite this specialization it will not be hard to see that our argument extends to general arithmetic situations.

1.1 Basic concepts

To begin with, we shall equip the upper half plane

$$\mathcal{H} = \{z = x + iy \ : \ -\infty < x < \infty, \ \ y > 0\}$$

with the non-Euclidean differentiable structure. For this purpose we introduce the group $\mathbb{T}(\mathcal{H})$ consisting of all real fractional linear transformations

$$\gamma : z \mapsto \frac{az+b}{lz+h} \qquad \big(ah - bl = 1;\ a,b,l,h \in \mathbb{R}\big). \tag{1.1.1}$$

The γ's map $\mathcal{H}$ onto itself conformally. To see this it is enough to note that γ has the inverse map $z \mapsto (hz-b)/(-lz+a)$, and that

$$\operatorname{Im}\gamma(z) = \frac{y}{|lz+h|^2}, \qquad \frac{d}{dz}\gamma(z) = \frac{1}{(lz+h)^2}. \tag{1.1.2}$$

The elements of $\mathbb{T}(\mathcal{H})$ are also rigid motions acting on $\mathcal{H}$ in the sense that $\mathcal{H}$ carries the non-Euclidean metric

$$y^{-1}|dz| = y^{-1}((dx)^2 + (dy)^2)^{\frac{1}{2}}$$

which is invariant with respect to any $\gamma \in \mathbb{T}(\mathcal{H})$. This is a simple consequence of the relations in (1.1.2), since they imply

$$(\operatorname{Im}\gamma(z))^{-1}\Big|\frac{d}{dz}\gamma(z)\Big| = (\operatorname{Im} z)^{-1}.$$

The invariance of the metric is inherited by Δ as being the negative of the corresponding Laplace–Beltrami operator. It can also be checked by direct computation: Putting $f(x, y) = F(u, v)$ with $\gamma(x + iy) = u + iv$ and invoking the Cauchy–Riemann equation for the function $\gamma(z)$, we have

$$\begin{aligned}\Delta f(x, y) &= -y^2(u_x^2 + v_x^2)(F_{uu} + F_{vv})\\ &= -y^2\Big|\frac{d}{dz}\gamma(z)\Big|^2(F_{uu} + F_{vv}) = -v^2(F_{uu} + F_{vv}),\end{aligned}$$

which amounts to $\Delta \cdot \gamma = \gamma \cdot \Delta$, i.e., the invariance of Δ. We have also the invariance of the non-Euclidean area element

$$d\mu(z) = y^{-2}dxdy$$

induced by the metric. This can be confirmed by computing the Jacobian of the map γ :

$$\Big|\frac{\partial(u, v)}{\partial(x, y)}\Big| = \Big|\frac{d}{dz}\gamma(z)\Big|^2 = (v/y)^2.$$

Further, we have the invariance of the non-Euclidean outer-normal derivative

$$y\frac{\partial}{\partial n} = y\Big\{\frac{dy}{|dz|}\frac{\partial}{\partial x} - \frac{dx}{|dz|}\frac{\partial}{\partial y}\Big\}$$

taken along any piecewise smooth curve in $\mathcal{H}$: In fact we have, for f, F as above,

$$\begin{aligned}y\frac{\partial f}{\partial n} &= y\Big\{\frac{dy}{|dz|}(F_u u_x + F_v v_x) - \frac{dx}{|dz|}(F_u u_y + F_v v_y)\Big\}\\ &= y\Big\{\frac{dy}{|dz|}(F_u v_y - F_v u_y) - \frac{dx}{|dz|}(-F_u v_x + F_v u_x)\Big\}\\ &= y\Big\{\frac{dv}{|dz|}F_u - \frac{du}{|dz|}F_v\Big\} = v\frac{\partial F}{\partial n}.\end{aligned}$$

This will be used in conjunction with Green's formula, which is a basic tool in the discussion below.

We next define the full modular group Γ as the subgroup of $\mathbb{T}(\mathcal{H})$ that is composed of those maps with $a, b, l, h \in \mathbb{Z}$ in (1.1.1). This signifies in particular that we do not regard Γ as a matrix group. Thus, if an element of Γ is pulled back to $SL(2, \mathbb{Z})$ in an obvious way, then we get two image matrices with corresponding entries having opposite signs; that is,

$$\Gamma \cong SL(2, \mathbb{Z})/\{\pm 1\}.$$

In any event readers should bear in mind that we are always dealing with transformations of $\mathcal{H}$.

The most basic fact about the motions caused by the elements of Γ is that they are discontinuous. This means that the action of Γ on $\mathcal{H}$ is comparable to, e.g., that of the group generated by two independent linear translations acting on the Euclidean plane, which is equivalent to tessellating $\mathbb{C}$ with congruent parallelograms. To make the situation with Γ explicit, we introduce the fundamental domain of Γ

$$\mathcal{F} = \left\{ z \in \mathcal{H} : |z| \geq 1, |x| \leq \tfrac{1}{2} \right\}, \tag{1.1.3}$$

and also the notation

$$z \equiv z' \bmod \Gamma$$

that indicates the existence of a $\gamma \in \Gamma$ such that $\gamma(z) = z'$. Then we have

Lemma 1.1 *The family of domains $\{\gamma(\mathcal{F}), \gamma \in \Gamma\}$ induces a tessellation of $\mathcal{H}$.*

Proof We fix an arbitrary $z \in \mathcal{H}$, and consider $\max[\mathrm{Im}\,\gamma(z)]$ as γ given in (1.1.1) varies in Γ. This should exist. For the first relation in (1.1.2) implies that $\mathrm{Im}\,\gamma(z)$ takes its maximum when $|lz + h|$ takes its minimum; and the latter can readily be seen to exist by observing that $lz + h$, $\gamma \in \Gamma$, are among the lattice points generated by 1 and z. We assume that $z_0 = x_0 + iy_0$ has the maximum imaginary part in this context; naturally we may assume also that $|x_0| \leq \frac{1}{2}$. Then we note that $-1/z_0 \equiv z_0 \bmod \Gamma$, and thus $\mathrm{Im}\,(-1/z_0) = y_0|z_0|^{-2} \leq y_0$. Hence we have $|z_0| \geq 1$, namely $z_0 \in \mathcal{F}$. This means that the tiles $\gamma(\mathcal{F})$, $\gamma \in \Gamma$, cover $\mathcal{H}$. We shall next show that these tiles have common points only on their boundaries. This is clearly equivalent to the assertion that if $z_1 = x_1 + iy_1$ and $\gamma(z_1) = x_2 + iy_2$ with a non-trivial $\gamma \in \Gamma$ are in $\mathcal{F}$ then both are on $\partial\mathcal{F}$, the boundary of $\mathcal{F}$. To see this let γ be as in (1.1.1) with integral coefficients. Obviously we may suppose also that $l \geq 0$ as well as $y_2 \geq y_1$. Comparing the imaginary parts of z_1 and $\gamma(z_1)$ we have $|lz_1+h|\le 1$, which implies $ly_1\le 1$. On noting that $y_1\ge \frac{1}{2}\sqrt{3}$, we find that l is equal either to 0 or to 1. If $l=0$ then $ah=1$; and $\gamma(z)=z\pm b$. Here $b=0$ is excluded because of an obvious reason. Thus $b=\pm 1$, and $z_1,\gamma(z_1)\in\partial\mathcal{F}$. On the other hand, if $l=1$ then $|z_1+h|\le 1$. Thus $|h|\le 1$. If $h=0$ then $|z_1|=1$ and $\gamma(z)=a-1/z$; and if $h=\pm 1$ then $z_1=\frac{1}{2}(\mp 1+i\sqrt{3})$ and $\gamma(z)=a-1/(z\pm 1)$. We readily get $|a|\le 1$, and hence $z_1,\gamma(z_1)\in\partial\mathcal{F}$ again. This ends the proof.

We note that tiles $\gamma(\mathcal{F})$, $\gamma\in\Gamma$, are generally different in shape for our Euclidean eyes, but if they were corrected with the metric $y^{-1}|dz|$ they would look just like each other. We remark also that the left and the right vertical edges of $\mathcal{F}$ are obviously equivalent to each other mod Γ, and that the circular part of $\partial\mathcal{F}$ is mapped onto itself by $z\mapsto -1/z$; thus the left half of the arc is equivalent to the right half. The identification of the equivalent boundary elements of $\mathcal{F}$ yields a punctured Riemann surface. The puncture corresponds to the point at infinity, and will be called the cusp of Γ in the sequel. The Riemann surface thus obtained is designated as the *manifold* $\mathcal{F}$, which carries the metric $y^{-1}|dz|$ with an obvious localization. Without this specification the symbol $\mathcal{F}$ stands for the fundamental domain of Γ. In passing we stress that the possible overlapping of $\gamma(\mathcal{F})$'s on their boundaries will not raise any pathological situations in our later discussion.

Turning to the analytical aspect, we introduce the concept of automorphy: A function f defined on $\mathcal{H}$ is said to be Γ-automorphic if $f(\gamma(z))=f(z)$ for all $\gamma\in\Gamma$. This is the same as to have f defined originally on the manifold $\mathcal{F}$ and to view it as a function over $\mathcal{H}$ in an obvious way. In this context the invariance of Δ means precisely that Δ can be regarded as a differential operator acting on the manifold $\mathcal{F}$.

A very important example of Γ-automorphic functions is the Poincaré series: We put, for a non-negative integer m and a complex number s,

$$P_m(z,s)=\sum_{\gamma\in\Gamma_\infty\backslash\Gamma}(\operatorname{Im}\gamma(z))^s e(m\gamma(z))\qquad \left(z\in\mathcal{H},\ \operatorname{Re}s>1\right),\tag{1.1.4}$$

where $e(z)=\exp(2\pi iz)$, Γ_∞ is the stabilizer in Γ of the cusp, i.e., the cyclic subgroup generated by the translation $z\mapsto z+1$, and γ runs over a representative set of the left cosets of Γ_∞ in Γ. The summands are independent of the choice of the representatives, and the sum converges absolutely, as can be seen from the expression (1.1.5) below. Hence $P_m(z,s)$ is Γ-automorphic. We note that the relation $\eta\gamma^{-1}\in\Gamma_\infty$, where $\gamma,\eta\in\Gamma,\gamma(z)=(az+b)/(lz+h)$, $\eta(z)=(a'z+b')/(l'z+h')$ with $l,l'\ge 0$, is equivalent either to $l = l'$, $h = h'$ or to $\gamma, \eta \in \Gamma_\infty$. Also we observe that for $l > 0$

$$\gamma(z) = \frac{a}{l} - \frac{1}{l(lz+h)}, \quad ah \equiv 1 \bmod l.$$

Thus we have

$$P_m(z,s) = y^s e(mz) + y^s \sum_{l=1}^{\infty} \sum_{\substack{h=-\infty \\ (h,l)=1}}^{\infty} |lz+h|^{-2s} e\Big(\frac{mh^*}{l}\Big) \exp\Big(-\frac{2\pi m i}{l(lz+h)}\Big) \tag{1.1.5}$$

with $hh^* \equiv 1 \bmod l$. We classify the summands according to $h \bmod l$, so that we have

$$P_m(z,s) = y^s e(mz) + y^s \sum_{l=1}^{\infty} l^{-2s} \sum_{\substack{h=1 \\ (h,l)=1}}^{l} e(mh^*/l)$$
$$\times \sum_{n=-\infty}^{\infty} |z + h/l + n|^{-2s} \exp\Big(-\frac{2\pi m i}{l^2(z+h/l+n)}\Big).$$

Applying Poisson's sum-formula to the last sum, we get, for $\operatorname{Re} s > 1$,

$$P_m(z,s) = y^s e(mz) + y^{1-s} \sum_{n=-\infty}^{\infty} e(nx) \sum_{l=1}^{\infty} l^{-2s} S(m,n\,;l)$$
$$\times \int_{-\infty}^{\infty} \exp\Big(-2\pi n y \xi i - \frac{2\pi m}{l^2 y(1-\xi i)}\Big)(1+\xi^2)^{-s} d\xi, \tag{1.1.6}$$

where

$$S(m,n;l) = \sum_{\substack{h=1 \\ (h,l)=1}}^{l} e((hm+h^*n)/l) \tag{1.1.7}$$

is the Kloosterman sum. Here we have performed an exchange of the order of summation, which is legitimate because the convergence is absolute, as can be seen by shifting the path to $\operatorname{Im}\xi = -\frac{1}{2}\operatorname{sgn}(n)$ in the integral. The formula (1.1.6) will be used in the next chapter.

In particular we put

$$E(z,s) = P_0(z,s) = \sum_{\gamma \in \Gamma_\infty \backslash \Gamma} (\operatorname{Im} \gamma(z))^s, \tag{1.1.8}$$

and call it the Eisenstein series attached to Γ. This function will appear at various important stages of our discussion. Its principal properties are collected in

Lemma 1.2 *For any $z \in \mathcal{H}$ the function $E(z,s)$ is meromorphic in s over the whole of $\mathbb{C}$, and we have the expansion*

$$E(z,s) = y^s + \varphi_\Gamma(s)y^{1-s} + \frac{2\pi^s\sqrt{y}}{\Gamma(s)\zeta(2s)}\sum_{n\neq 0}|n|^{s-\frac{1}{2}}\sigma_{1-2s}(|n|)K_{s-\frac{1}{2}}(2\pi|n|y)e(nx), \tag{1.1.9}$$

where

$$\varphi_\Gamma(s) = \sqrt{\pi}\frac{\Gamma(s-\frac{1}{2})\zeta(2s-1)}{\Gamma(s)\zeta(2s)} = \frac{\pi^{2s-1}\Gamma(1-s)\zeta(2(1-s))}{\Gamma(s)\zeta(2s)}. \tag{1.1.10}$$

Hence $E(z,s)$ is regular for $\operatorname{Re} s \geq \frac{1}{2}$ *save for the simple pole at $s=1$ with residue $3/\pi$, and satisfies the functional equation*

$$E(z,s) = \varphi_\Gamma(s)E(z,1-s) \tag{1.1.11}$$

as well as the differential equation

$$\Delta E(z,s) = s(1-s)E(z,s). \tag{1.1.12}$$

Proof We invoke first the functional equation

$$\zeta(s) = 2^s\pi^{s-1}\sin(\tfrac{1}{2}s\pi)\Gamma(1-s)\zeta(1-s) = \pi^{s-\frac{1}{2}}\frac{\Gamma(\frac{1}{2}(1-s))}{\Gamma(\frac{1}{2}s)}\zeta(1-s) \tag{1.1.13}$$

and the representation

$$\sigma_\xi(n) = \zeta(1-\xi)\sum_{l=1}^{\infty} l^{\xi-1}c_l(n) \quad (n>0,\ \operatorname{Re}\xi<0), \tag{1.1.14}$$

where

$$c_l(n) = S(n,0;l) = \sum_{\substack{h=1\\(h,l)=1}}^{l} e(nh/l) \tag{1.1.15}$$

is the Ramanujan sum. When $\operatorname{Re} s > 1$ the expansion (1.1.9) is a consequence of the relations (1.1.6) and (1.1.14) with (1.1.13), since we have

$$\int_{-\infty}^{\infty}\frac{1}{(1+\xi^2)^{\nu+\frac{1}{2}}}d\xi = \sqrt{\pi}\frac{\Gamma(\nu)}{\Gamma(\nu+\frac{1}{2})} \qquad (\operatorname{Re}\nu>0)$$

and

$$\int_{-\infty}^{\infty}\frac{\cos(y\xi)}{(1+\xi^2)^{\nu+\frac{1}{2}}}d\xi = 2\sqrt{\pi}\frac{(y/2)^\nu}{\Gamma(\nu+\frac{1}{2})}K_\nu(y) \qquad (y>0;\ \operatorname{Re}\nu>-\tfrac{1}{2}). \tag{1.1.16}$$

We then note the other representation

$$K_\nu(y) = \frac{1}{2}\int_0^\infty \xi^{\nu-1}\exp\left(-\tfrac{1}{2}y(\xi+\xi^{-1})\right)d\xi, \tag{1.1.17}$$

which shows that $K_\nu(y)$ is entire in ν, $K_\nu(y) = K_{-\nu}(y)$ for $\operatorname{Re} y > 0$, and moreover

$$K_\nu(y) = (1+o(1))\left(\frac{\pi}{2y}\right)^{\frac{1}{2}} e^{-y} \qquad (y \to +\infty) \tag{1.1.18}$$

for any fixed ν. In fact this asymptotic formula can readily be proved by putting $\xi = 1 + r$ in (1.1.17) and observing that the main part of the integral comes from the short interval $|r| \le y^{-2/5}$. Hence $E(z,s)$ exists as a meromorphic function of s over $\mathbb{C}$. The assertion (1.1.11) follows from (1.1.13) and these properties of $K_\nu(y)$. As to (1.1.12) it is a consequence of the definition (1.1.8) and

$$\Delta[(\operatorname{Im}\gamma(z))^s] = s(1-s)(\operatorname{Im}\gamma(z))^s \qquad (s \in \mathbb{C},\ \gamma \in \mathbb{T}(\mathcal{H})), \tag{1.1.19}$$

which is due to the invariance of Δ. An alternative proof is to use the expansion (1.1.9) and the fact that $\sqrt{y}K_{s-\frac{1}{2}}(2\pi|n|y)$ is a solution of the differential equation

$$[D_{s,n}g](y) \equiv -y^2g''(y) + ((2\pi n y)^2 + s(s-1))g(y) = 0 \quad (y > 0) \tag{1.1.20}$$

(see the next lemma). This ends the proof of the lemma.

It is appropriate to make here a little digression about the nature of the differential operator $D_{s,n}$: It is a result of the application of the separation of variables to the operator $\Delta + s(s-1)$. In fact we have formally

$$(\Delta + s(s-1))\Big\{\sum_n a_n(y)e(nx)\Big\} = \sum_n [D_{s,n}a_n](y)e(nx). \tag{1.1.21}$$

This relation and the following assertion will be used in our later discussion.

Lemma 1.3 *The differential equation* (1.1.20) *with $n > 0$ has linearly independent solutions $\sqrt{y}K_{s-\frac{1}{2}}(2\pi n y)$ and $\sqrt{y}I_{s-\frac{1}{2}}(2\pi n y)$. Thus the resolvent kernel of the differential operator $y^{-2}D_{s,n}$, $n > 0$, is equal to*

$$g_{s,n}(y,v) = \begin{cases} \sqrt{vy}\,I_{s-\frac{1}{2}}(2\pi n v)K_{s-\frac{1}{2}}(2\pi n y) & \text{if } v \le y, \\[2ex] \sqrt{vy}\,I_{s-\frac{1}{2}}(2\pi n y)K_{s-\frac{1}{2}}(2\pi n v) & \text{if } v \ge y. \end{cases} \tag{1.1.22}$$

Proof That the above two functions are solutions of (1.1.20), and that the Wronskian of these is equal to 1, can be checked by using the recurrence relations

$$I_{\nu-1}(z)+I_{\nu+1}(z)=2I_\nu'(z),\quad I_{\nu-1}(z)-I_{\nu+1}(z)=2\nu z^{-1}I_\nu(z),$$
$$K_{\nu-1}(z)+K_{\nu+1}(z)=-2K_\nu'(z),\quad K_{\nu-1}(z)-K_{\nu+1}(z)=-2\nu z^{-1}K_\nu(z), \tag{1.1.23}$$

and also the definitions

$$I_\nu(z)=\sum_{k=0}^{\infty}\frac{(z/2)^{\nu+2k}}{\Gamma(k+1)\Gamma(\nu+k+1)}, \tag{1.1.24}$$

$$K_\nu(z)=\frac{\pi}{2\sin(\pi\nu)}\{I_{-\nu}(z)-I_\nu(z)\}, \tag{1.1.25}$$

where $z^\nu = \exp(\nu\log z)$ with $|\arg z| < \pi$. The rest of the proof is a standard application of the general theory of ordinary differential equations. The excluded case $n = 0$ is easy, and left for readers. This ends the proof.

Proceeding to our main issue, we let $L^2(\mathcal{F}, d\mu)$ stand for the set of all Γ-automorphic f's such that

$$\|f\|^2=\int_{\mathcal{F}}|f(z)|^2d\mu(z)<+\infty.$$

It should be observed that $\mathbb{C}\subset L^2(\mathcal{F}, d\mu)$, since we have

$$\int_{\mathcal{F}}d\mu(z)=\frac{\pi}{3}.$$

The set $L^2(\mathcal{F}, d\mu)$ is a Hilbert space equipped with the Petersson inner-product

$$\langle f_1,f_2\rangle=\int_{\mathcal{F}}f_1(z)\overline{f_2(z)}d\mu(z). \tag{1.1.26}$$

We are going to diagonalize the operator Δ in $L^2(\mathcal{F}, d\mu)$; that is, we shall try to find a set of Γ-automorphic functions which spans $L^2(\mathcal{F}, d\mu)$ and in which the operator Δ is well-defined and reduces, in an informal sense, to a scalar multiplication at each element. To this end we introduce the linear set defined on the manifold $\mathcal{F}$:

$$B^\infty(\mathcal{F})=\{f\in C^\infty(\mathcal{F}):\ \text{each partial derivative of } f(z) \text{ is of rapid decay}\}. \tag{1.1.27}$$

Here, that a Γ-automorphic function $g(z)$ is of rapid decay means that

$g(z) = O(y^{-M})$ for any $M > 0$ as $z \in \mathcal{F}$ tends to the cusp, where the implied constant may depend on M. This is dense in $L^2(\mathcal{F}, d\mu)$, for it contains all C^∞-functions having compact supports on the manifold $\mathcal{F}$. Then Δ is a symmetric operator in the sense that

$$\langle \Delta f_1, f_2 \rangle = \langle f_1, \Delta f_2 \rangle \qquad (f_1, f_2 \in B^\infty(\mathcal{F})). \tag{1.1.28}$$

In fact we have, by Green's formula,

$$\langle \Delta f_1, f_2 \rangle = \int_{\mathcal{F}} y^2 \nabla f_1(z) \cdot \overline{\nabla f_2(z)} d\mu(z), \tag{1.1.29}$$

where ∇ denotes the ordinary gradient. For the manifold $\mathcal{F}$ has no boundary, and the integrand is of rapid decay around the cusp. More precisely, we apply Green's formula to

$$\langle \Delta f_1, f_2 \rangle = \lim_{Y \to \infty} \int_{\mathcal{F}_Y} \Delta f_1(z) \overline{f_2(z)} d\mu(z), \tag{1.1.30}$$

where $\mathcal{F}_Y = \mathcal{F} \cap \{z : \operatorname{Im} z \le Y\}$. We have

$$\begin{aligned} \langle \Delta f_1, f_2 \rangle &= \lim_{Y \to \infty} \int_{\mathcal{F}_Y} \nabla f_1(z) \cdot \overline{\nabla f_2(z)}\, dxdy \\ &\quad - \lim_{Y \to \infty} \int_{\partial \mathcal{F}_Y} y \frac{\partial f_1}{\partial n}(z) \overline{f_2(z)} \frac{|dz|}{y} \\ &= \int_{\mathcal{F}} \nabla f_1(z) \cdot \overline{\nabla f_2(z)}\, dxdy \\ &\quad - \lim_{Y \to \infty} \int_{-\frac{1}{2}}^{\frac{1}{2}} \frac{\partial f_1}{\partial y}(x + iY) \overline{f_2(x + iY)} dx, \end{aligned} \tag{1.1.31}$$

which gives (1.1.29). Here we have used the invariance of $(\partial f_1 / \partial n) \overline{f_2} |dz|$ and the consequential cancellation of integrals on the boundary elements of $\mathcal{F}$ which are equivalent $\bmod \Gamma$ to each other. This cancellation is due to the invariance of the non-Euclidean outer-normal derivative, and to the reverse of orientation in the corresponding boundary elements. Intuitively it is the same as what happens when the process of folding and pasting is applied to $\mathcal{F}$ to transform it into a Riemann surface.

The formula (1.1.29) implies, in particular,

$$\langle \Delta f, f \rangle > 0 \tag{1.1.32}$$

for any non-constant $f \in B^\infty(\mathcal{F})$. This and (1.1.28) mean that Δ is a semi-bounded symmetric operator which has a dense domain in $L^2(\mathcal{F}, d\mu)$. Thus we could appeal to the general theory on the self-adjoint extension of such an operator with the effect of a shorter presentation. We shall, however, dispense with the operator theory in order to make our discussion as elementary as possible.

We shall next show that $B^\infty(\mathcal{F})$ is a proper place to look for L^2-eigenfunctions of Δ:

Lemma 1.4 *Let $f \in C^2(\mathcal{F}) \cap L^2(\mathcal{F}, d\mu)$ be such that $\Delta f = (\frac{1}{4} + \kappa^2)f$ with $\operatorname{Im} \kappa \geq 0$, $\kappa \neq \frac{1}{2}i$. Then we have the absolutely convergent expansion*

$$f(z) = y^{\frac{1}{2}} \sum_{n \neq 0} \rho(n) K_{i\kappa}(2\pi|n|y)e(nx) \quad (z \in \mathcal{H}) \tag{1.1.33}$$

with certain complex numbers $\rho(n)$. Thus $f \in B^\infty(\mathcal{F})$; and this implies further that

$$\kappa > 3.815. \tag{1.1.34}$$

Proof As is shown below, f is a constant if $\kappa = \frac{1}{2}i$. Thus this case is excluded in the above; and we may assume that f is non-trivial. Since f is of period 1 in the variable x, it can be expanded into an absolutely convergent Fourier series

$$f(z) = \sum_{n=-\infty}^{\infty} a(n, y)e(nx) \qquad (z \in \mathcal{H}). \tag{1.1.35}$$

Expressing $a(n, y)$ in terms of the integral of $f(x + iy)e(-nx)$ over the unit interval and applying integration by parts appropriately, we get

$$D_{\frac{1}{2}+i\kappa, n} a(n, y) = 0$$

(cf. (1.1.21)). Thus we see that

$$a(0, y) = \begin{cases} c_0 y^{\frac{1}{2}+i\kappa} + d_0 y^{\frac{1}{2}-i\kappa} & \text{if } \kappa \neq 0, \\ c_0 y^{\frac{1}{2}} + d_0 y^{\frac{1}{2}} \log y & \text{if } \kappa = 0, \end{cases}$$

and by Lemma 1.3 that if $n \neq 0$

$$a(n, y) = c_n y^{\frac{1}{2}} K_{i\kappa}(2\pi|n|y) + d_n y^{\frac{1}{2}} I_{i\kappa}(2\pi|n|y).$$

On the other hand the condition $f \in L^2(\mathcal{F}, d\mu)$ gives

$$\sum_{n=-\infty}^{\infty} \int_1^\infty |a(n, y)|^2 y^{-2} dy \leq \int_{\mathcal{F}} |f(z)|^2 d\mu(z) < \infty. \tag{1.1.36}$$

Taking account of the fact that K_ν is of exponential decay (see (1.1.18)) and I_ν of exponential growth (see (2.6.23)), we find that $d_n = 0$ for $n \neq 0$, and hence

$$f(z) = a(0,y) + y^{\frac{1}{2}} \sum_{n \neq 0} c_n K_{i\kappa}(2\pi|n|y)e(nx). \tag{1.1.37}$$

Here it should be observed that we have, for any fixed $\delta > 0$,

$$c_n = O(e^{\delta|n|}). \tag{1.1.38}$$

For the series in (1.1.37) is absolutely convergent for any given $y > 0$ and we have the asymptotic formula (1.1.18). We then need to prove that $a(0,y) \equiv 0$. This is obvious if κ is real. To deal with the case $\operatorname{Im}\kappa > 0$ we consider the trivial equation

$$0 = \int_{\mathcal{F}_Y} \left[E(z,s)\Delta f(z) - f(z)\Delta E(z,s)\right] d\mu(z) \quad (s = \tfrac{1}{2} - i\kappa \neq 1),$$

where $\mathcal{F}_Y$ is as above but with a sufficiently large Y, and we have used (1.1.12). In much the same way as in (1.1.31), Green's formula reduces this to

$$0 = -\int_{-\frac{1}{2}}^{\frac{1}{2}} \left[E(z,s)\frac{\partial}{\partial y}f(z) - f(z)\frac{\partial}{\partial y}E(z,s)\right] dx \qquad (\operatorname{Im} z = Y).$$

We insert (1.1.9) and (1.1.37) into the right side and let Y tend to $+\infty$, getting

$$c_0 = d_0 \varphi_\Gamma(\tfrac{1}{2} - i\kappa), \tag{1.1.39}$$

where we have used (1.1.38). We note that $\varphi_\Gamma(\frac{1}{2} - i\kappa)$ is finite. But we have $d_0 = 0$ if $\operatorname{Im}\kappa > 0$, because of (1.1.36); and thus we have $c_0 = 0$ too. This ends the proof of (1.1.33). Then the fact that $f \in B^\infty(\mathcal{F})$ follows immediately.

We next proceed to the proof of (1.1.34). For this purpose we observe that the integrand in (1.1.29) is invariant under the action of Γ, which can be proved just as that of Δ. Thus we may replace the integration domain $\mathcal{F}$ by any of $\gamma(\mathcal{F})$, $\gamma \in \Gamma$. We put $f_1 = f_2 = f$ with f as above, and use two domains, $\mathcal{F}$ and its image $\mathcal{F}^*$ under the mapping $z \mapsto -1/z$. We have

$$2(\kappa^2 + \tfrac{1}{4})\|f\|^2 = \int_{\mathcal{F}\cup\mathcal{F}^*} |\nabla f|^2 dxdy \geq \int_{y_0}^{\infty} \int_{-\frac{1}{2}}^{\frac{1}{2}} |\partial f/\partial x|^2 dxdy,$$

where $y_0 = \frac{1}{2}\sqrt{3}$. We insert the expansion (1.1.35) into this. We have

$$2(\kappa^2 + \tfrac{1}{4})\|f\|^2 \geq 4\pi^2 \int_{y_0}^{\infty} \sum_n n^2 |a(n,y)|^2 dy$$
$$\geq 4\pi^2 y_0^2 \int_{y_0}^{\infty} \sum_{n\neq 0} |a(n,y)|^2 y^{-2} dy.$$

Then we observe that the last integral is not less than $\|f\|^2$, for we now have $a(0,y) \equiv 0$. Hence we get $\kappa^2 + \frac{1}{4} \geq \frac{3}{2}\pi^2$ provided f is non-trivial. This yields (1.1.34).

We consider the remaining case $\kappa = \frac{1}{2}i$. The above argument preceding (1.1.37) gives the expansion

$$f(z) = \sum_{n=-\infty}^{\infty} c_n e^{-2\pi|n|y + 2\pi i n x}.$$

Hence we may use (1.1.29) with $f_1 = f_2 = f - c_0$. Then we get $\nabla f \equiv 0$, or $f \equiv c_0$, as has been claimed above. This ends the proof of the lemma.

Now, combining the assertions verified in the above and anticipating what we shall prove rigorously below, we indicate briefly the structure of $L^2(\mathcal{F}, d\mu)$: The set of eigenvalues of Δ corresponding to those eigenfunctions in $B^\infty(\mathcal{F})$ will turn out to be an unbounded discrete subset of the half line $[\frac{3}{2}\pi^2, \infty)$; the infinitude of the cardinality of this set is to be proved not in the present but in the next chapter (see (2.3.14)). We designate these eigenvalues as

$$\lambda_j = \kappa_j^2 + \tfrac{1}{4} \quad (\kappa_j > 0,\ j = 1, 2, \ldots) \tag{1.1.40}$$

in non-decreasing order; note that we count them including their multiplicities, which are always finite. The set $\{\lambda_j : j \geq 1\} \cup \{0\}$ is called the discrete spectrum of Δ with respect to Γ, or just the spectrum of Γ. Let ψ_j belong to the eigen-space corresponding to λ_j. We call it either a Maass wave or a real-analytic cusp-form over Γ. According to (1.1.33) we have the expansion

$$\psi_j(z) = y^{\frac{1}{2}} \sum_{n\neq 0} \rho_j(n) K_{i\kappa_j}(2\pi|n|y) e(nx) \quad (z \in \mathcal{H}) \tag{1.1.41}$$

with certain complex numbers $\rho_j(n)$. We note that if $\lambda_j \neq \lambda_{j'}$ then ψ_j and

$\psi_{j'}$ are orthogonal. For we have, by Green's formula,

$$\begin{aligned}(\lambda_j - \lambda_{j'})\langle\psi_j, \psi_{j'}\rangle &= \int_{\mathcal{F}} \left[(\Delta\psi_j)(z)\overline{\psi_{j'}(z)} - \psi_j(z)\overline{(\Delta\psi_{j'})(z)}\,\right] d\mu(z) \\ &= -\int_{\partial\mathcal{F}} \left[y(\partial\psi_j/\partial n)(z)\overline{\psi_{j'}(z)}\right. \\ &\qquad \left. - \psi_j(z)\overline{y(\partial\psi_{j'}/\partial n)(z)}\,\right]\frac{|dz|}{y} = 0. \end{aligned} \tag{1.1.42}$$

Thus we may assume naturally that

$$\{\psi_j : j \geq 1\} \text{ is an orthonormal system.} \tag{1.1.43}$$

Then $\rho_j(n)$'s are called Maass–Fourier coefficients, and will play a fundamental rôle throughout this monograph.

The difficulty of the harmonic analysis over the manifold $\mathcal{F}$ lies in the fact that the closure of the linear set generated by the system (1.1.43) does not fill the whole of $L^2(\mathcal{F}, d\mu)$, although this is expected because the manifold $\mathcal{F}$ is not compact, i.e., punctured. The orthogonal complementary subspace is fairly large; and what is remarkable is that it can be completely described in terms of the Eisenstein series, as is embodied in

Theorem 1.1 *The set of eigenvalues of the non-Euclidean Laplacian Δ acting over $B^\infty(\mathcal{F})$ is discrete. Accordingly we have the orthogonal decomposition*

$$L^2(\mathcal{F}, d\mu) = \mathbb{C} \oplus \mathfrak{D} \oplus \mathfrak{E}. \tag{1.1.44}$$

Here $\mathfrak{D}$ is spanned by the system (1.1.43), *and*

$$\mathfrak{E} = \left\{ \operatorname*{l.i.m.}_{T\to\infty} \int_0^T p(t)E(z, \tfrac{1}{2} + it)dt \; : \; p \in L^2(0,\infty)\right\},$$

where the limit is taken in the space $L^2(\mathcal{F}, d\mu)$, and $L^2(0,\infty)$ is the ordinary Hilbert space composed of all functions which are square-integrable over the positive real axis with respect to the Lebesgue measure. More precisely, we have, for any $f \in L^2(\mathcal{F}, d\mu)$,

$$f = \sum_{j=0}^{\infty} \langle f, \psi_j\rangle\psi_j + \frac{1}{2\pi}\int_0^\infty \mathcal{E}(t, f)E(z, \tfrac{1}{2} + it)dt \tag{1.1.45}$$

in the sense of the norm convergence. Here $\psi_0 \equiv (3/\pi)^{\frac{1}{2}}$ and

$$\mathcal{E}(t, f) = \operatorname*{l.i.m.}_{Y\to\infty} \int_{\mathcal{F}_Y} f(z)E(z, \tfrac{1}{2} - it)\, d\mu(z), \tag{1.1.46}$$

where $\mathcal{F}_Y$ is as in (1.1.30), *and the limit is taken in the space* $L^2(0,\infty)$. *Thus we have the Parseval formula: For any* $f_1, f_2 \in L^2(\mathcal{F}, d\mu)$

$$\langle f_1, f_2\rangle = \sum_{j=0}^{\infty}\langle f_1, \psi_j\rangle\overline{\langle f_2, \psi_j\rangle} + \frac{1}{2\pi}\int_0^\infty \mathcal{E}(t, f_1)\overline{\mathcal{E}(t, f_2)}dt. \tag{1.1.47}$$

We shall develop a proof of this fundamental theorem in the subsequent sections; the existence of the above limits in mean will be established at the final stage of our discussion. It should be remarked that the functional equation (1.1.11) implies that

$$\mathcal{E}(t, f) = \varphi_\Gamma(\tfrac{1}{2} + it)\mathcal{E}(-t, f),$$

and thus $|\mathcal{E}(t,f)|$, $\mathcal{E}(t,f)E(z,\frac{1}{2}+it)$, $\mathcal{E}(t,f_1)\overline{\mathcal{E}(t,f_2)}$ are all even functions of the real variable t. Hence each integral in (1.1.45) and (1.1.47) can be replaced by the half of the one over the line $(-\infty,\infty)$, as we shall do in our later discussion.

Here we describe briefly the salient points of our proof: Our problem is equivalent to proving the discreteness of the eigenvalues of Δ while gaining the control of the continuous spectrum that is expected to occur. To this end we consider the orthogonal decomposition

$$L^2(\mathcal{F}, d\mu) = \mathbb{C} \oplus \mathfrak{A} \oplus \mathfrak{B}, \tag{1.1.48}$$

where

$\mathfrak{A}$ = the closure of the linear set generated by all eigenfunctions of Δ found in $B^\infty(\mathcal{F})$,

and $\mathfrak{B}$ is the orthogonal complement of $\mathbb{C}\oplus\mathfrak{A}$. That the constant 1 is orthogonal to $\mathfrak{A}$ is immediate (cf. (1.1.42)). It should be noted that we do not take for granted that the dimension of $\mathfrak{A}$ is infinite; that will be proved in the next chapter, as we have mentioned already. We are going to prove that $\mathfrak{A} = \mathfrak{D}$ and $\mathfrak{B} = \mathfrak{C}$. The key idea is to translate the problem into a situation which can be dealt with by the Hilbert–Schmidt theory on integral transformations with bounded continuous symmetric kernels. The relevant integral transformation should have a dense image of $\mathfrak{A}$ in itself without moving each eigen-space of Δ wildly. Because of the definition of $\mathfrak{A}$ a possible choice is the resolvent of the differential operator $\Delta + \lambda$ on $\mathcal{F}$, for it is a non-zero scalar multiplication at each generating element of $\mathfrak{A}$, provided $|\lambda|$ is not too small. Hence we shall consider first the *free-space* resolvent kernel $r_\alpha(z, w)$ of $\Delta + \alpha(\alpha - 1)$ on $\mathcal{H}$, where α is assumed to be real for the sake of simplicity. It will be shown that

$$r_\alpha(z,w) = \frac{\Gamma(\alpha)^2}{4\pi\Gamma(2\alpha)}\left(\left|\frac{z-\overline{w}}{z-w}\right|^2 - 1\right)^\alpha F\left(\alpha,\alpha;2\alpha;1-\left|\frac{z-\overline{w}}{z-w}\right|^2\right), \quad (1.1.49)$$

where F is the hypergeometric function (see (1.2.7)). We shall then consider the transformation

$$\mathcal{R}_\alpha f(z) = \int_{\mathcal{H}} r_\alpha(z,w)f(w)d\mu(w).$$

When f is Γ-automorphic, this becomes

$$\mathcal{R}_\alpha f(z) = \int_{\mathcal{F}} R_\alpha(z,w)f(w)d\mu(w), \quad (1.1.50)$$

where

$$R_\alpha(z,w) = \sum_{\gamma\in\Gamma} r_\alpha(z,\gamma(w)), \quad (1.1.51)$$

which will turn out to be absolutely convergent for $\alpha > 1$, $z \not\equiv w \bmod \Gamma$. It is known that $\mathcal{R}_\alpha$ is the resolvent of $\Delta + \alpha(\alpha-1)$ (see Section 1.5). But in our argument we shall need only the fact that $\mathcal{R}_\alpha$ is the *left-inverse* of $\Delta + \alpha(\alpha-1)$, i.e.,

$$\mathcal{R}_\alpha(\Delta + \alpha(\alpha-1))f = f \quad (1.1.52)$$

for any sufficiently smooth f (see (1.2.11)). Thus $\mathcal{R}_\alpha$ reduces to a non-vanishing scalar multiplication at each generating element of $\mathfrak{A}$: If $\Delta f = \lambda f$ then the above implies

$$\mathcal{R}_\alpha f = (\lambda + \alpha(\alpha-1))^{-1} f$$

provided α is sufficiently large. Hence it can be a candidate for the integral transformation mentioned above. We have, however, two difficulties with the kernel R_α. It is neither continuous nor bounded. The former is obviously inevitable; and a little investigation reveals that the latter is due to the fact that (1.1.51) contains the infinitely winding part corresponding to $\gamma \in \Gamma_\infty$. In order to eliminate the discontinuity, we consider

$$\mathcal{R}_{\alpha,1} = (2\alpha)^{-1}\{\mathcal{R}_\alpha - \mathcal{R}_{\alpha+1}\}.$$

For the corresponding kernel

$$R_{\alpha,1}(z,w) = (2\alpha)^{-1}\{R_\alpha(z,w) - R_{\alpha+1}(z,w)\}$$

is continuous, and at the same time $\mathcal{R}_{\alpha,1}$ is essentially diagonal over $\mathfrak{A}$. But $R_{\alpha,1}$ is still unbounded. So, to get a compactification of $\mathcal{R}_{\alpha,1}$, we try to subtract from $R_{\alpha,1}$ a certain symmetric Γ-automorphic kernel $C_{\alpha,1}$ of simple nature. The asymptotic character of $C_{\alpha,1}$ should naturally be the same as that of $R_{\alpha,1}$; and moreover this modification of $R_{\alpha,1}$ should not alter the action of $\mathcal{R}_{\alpha,1}$ too much. It is optimal to have a $C_{\alpha,1}$ which annihilates $\mathfrak{A}$. Thus the asymptotic evaluation of the contribution of Γ_∞ to $R_{\alpha,1}$ is to be developed. The main term will prove to be equal to

$$\frac{1}{2\pi}\int_{-\infty}^{\infty} q_{\alpha,1}(t^2+\tfrac{1}{4})(\operatorname{Im} w)^{\frac{1}{2}+it}(\operatorname{Im} z)^{\frac{1}{2}-it}dt,$$

and the rest is bounded over $\mathcal{F}\times\mathcal{F}$, where

$$q_{\alpha,1}(\lambda) = \left[(\lambda+\alpha(\alpha-1))(\lambda+\alpha(\alpha+1))\right]^{-1}.$$

In view of (1.1.9) this result suggests that

$$C_{\alpha,1}(z,w) = \frac{1}{4\pi}\int_{-\infty}^{\infty} q_{\alpha,1}(t^2+\tfrac{1}{4})E(z,\tfrac{1}{2}+it)E(w,\tfrac{1}{2}-it)dt$$

would be the right choice. We can show that this is indeed the case. Then the discreteness of the eigenvalues of Δ is a simple consequence of Bessel's inequality, which amounts to $\mathfrak{A}=\mathfrak{D}$. It is also visible in the last expression that the Eisenstein series is the key to understanding the continuous spectrum.

The rest of the proof of the theorem is devoted to the verification of the Parseval formula (1.1.47), i.e., the separation of the continuous spectrum, which will be achieved by virtue of the classical result of Hilbert and Schmidt on the existence of eigenvalues of integral transformations. However, at this stage we would encounter a difficulty pertaining to convergence if we continued to use the transformation $\mathcal{R}_{\alpha,1}$. In order to avoid it we observe that (1.1.52) gives

$$\mathcal{R}_{\alpha,1}(\Delta+\alpha(\alpha-1))(\Delta+\alpha(\alpha+1))f = f.$$

Hence $R_{\alpha,1}$ is the iterated kernel coming from R_α and $R_{\alpha+1}$. Since it is a general practice to consider higher iteration if smoother expressions are required, we shall work with the integral transformation $\mathcal{R}_{\alpha,N}$ such that

$$\mathcal{R}_{\alpha,N}\prod_{\nu=0}^{N}(\Delta+(\alpha+\nu)(\alpha+\nu-1))f = f.$$

The kernel corresponding to $\mathcal{R}_{\alpha,N}$ is a linear combination of $R_{\alpha+\nu}$, $0\le\nu\le N$, which generalizes the above expression for $R_{\alpha,1}$. We shall show that $\mathcal{R}_{\alpha,N}$ with a relatively small N provides us with ample control of convergence while retaining the basic properties of $\mathcal{R}_{\alpha,1}$ indicated above.

A lower bound of N will be given, although it is obviously immaterial to our principal purpose.

1.2 Resolvent kernel

We now start our proof of the spectral resolution (1.1.44) of Δ over the full modular group Γ. To this end we shall first study basic properties of the kernel $r_\alpha(z,w)$. We note that the real parameter α can be assumed to be not less than $\frac{1}{2}$ because of symmetry.

We have, by the Mellin–Barnes formula for the hypergeometric function,

$$r_\alpha(z,w) = p_\alpha(\varrho(z,w)); \tag{1.2.1}$$

$$p_\alpha(\varrho) = \frac{1}{8\pi^2 i}\int_{(-\frac{1}{4})} \frac{\Gamma(\xi+\alpha)^2\Gamma(-\xi)}{\Gamma(\xi+2\alpha)}\varrho^{-\xi-\alpha}d\xi, \tag{1.2.2}$$

where the integral is taken over the vertical line $\mathrm{Re}\,\xi = -\frac{1}{4}$, and

$$\varrho(z,w) = \frac{|z-w|^2}{4(\mathrm{Im}\,z)(\mathrm{Im}\,w)} \qquad (z,w\in\mathcal{H}).$$

It should be noted that we have $\varrho(\gamma(z),\gamma(w)) = \varrho(z,w)$ for any $\gamma\in\mathbb{T}(\mathcal{H})$, which we shall use in the sequel without mention. That $r_\alpha(z,w)$ is the resolvent kernel of the operator $\Delta+\alpha(\alpha-1)$ over $\mathcal{H}$ is equivalent to

Lemma 1.5 *We have, for* $\alpha\ge\frac{1}{2}$,

$$(\Delta+\alpha(\alpha-1))_w r_\alpha(z,w) = 0 \qquad (z\ne w) \tag{1.2.3}$$

as well as

$$\left.\begin{aligned} p_\alpha(\varrho) &= -(4\pi)^{-1}\log\varrho + O(1) && (\varrho\to+0),\\ p_\alpha'(\varrho) &= -(4\pi\varrho)^{-1} + (4\pi)^{-1}\alpha(1-\alpha)\log\varrho + O(1) && (\varrho\to+0),\\ p_\alpha(\varrho) &= O(\varrho^{-\alpha}) && (\varrho\to+\infty). \end{aligned}\right\} \tag{1.2.4}$$

Proof To prove the first assertion we note that

$$\Delta_w p_\alpha(\varrho(z,w)) = -(\varrho^2+\varrho)p_\alpha''(\varrho) - (1+2\varrho)p_\alpha'(\varrho), \tag{1.2.5}$$

which follows from

$$\Delta_w p_\alpha(\varrho(z,w)) = -v^2\{p_\alpha''(\varrho)[\varrho_u^2+\varrho_v^2] + p_\alpha'(\varrho)[\varrho_{uu}+\varrho_{vv}]\} \qquad (w = u+vi),$$

and

$$\left.\begin{aligned} \varrho_u &= (u-x)/(2vy), \quad \varrho_v = (v^2-y^2-(u-x)^2)/(4v^2y), \\ \varrho_{uu} &= 1/(2vy), \qquad\quad \varrho_{vv} = (y^2+(u-x)^2)/(2v^3y). \end{aligned}\right\} \tag{1.2.6}$$

Then by (1.2.2) we have

$$\Delta_w p_\alpha(\varrho(z,w)) = -\frac{1}{8\pi^2 i}\int_{(-\frac{1}{4})} \frac{\Gamma(\xi+\alpha)^2\Gamma(-\xi)}{\Gamma(\xi+2\alpha)} \times \left\{(\xi+\alpha)(\xi+\alpha-1)\varrho^{-\xi-\alpha} + (\xi+\alpha)^2\varrho^{-\xi-\alpha-1}\right\}d\xi.$$

This implies a division of the right side into two parts in an obvious way. In the part involving $\varrho^{-\xi-\alpha-1}$ we shift the path to $\operatorname{Re}\xi = -\frac{5}{4}$ and replace ξ by $\xi - 1$. After a simple rearrangement we end the proof of (1.2.3). Then, to prove the first assertion in (1.2.4), we shift the path in (1.2.2) to $\operatorname{Re}\xi = -\alpha - \frac{1}{2}$; and to prove the second we differentiate (1.2.2) inside the integral, and shift the path to $\operatorname{Re}\xi = -\alpha - \frac{3}{2}$. As to the third assertion we need only to shift the path to $\operatorname{Re}\xi = \frac{1}{2}$. This ends the proof of the lemma.

We have yet to explain how the kernel $r_\alpha(z,w)$ has been fixed: As is suggested by the ordinary potential theory over the Euclidean plane, $r_\alpha(z,w)$ is to depend only on the non-Euclidean distance between z and w. On the other hand it is easy to see that this distance is a function of $\varrho(z,w)$. Thus we have $r_\alpha(z,w) = f(\varrho(z,w))$ with a certain f. Then, in view of (1.2.5), the equation (1.2.3) which the resolvent kernel has to satisfy can be translated into

$$\left[(\rho^2+\rho)(d/d\rho)^2 + (1+2\rho)(d/d\rho) + \alpha(1-\alpha)\right]f(\rho) = 0. \tag{1.2.7}$$

Because of the definition the desirable solution f should have a prescribed logarithmic singularity at $\rho = 0$; hence we have $f = p_\alpha$. The assertions (1.2.1)–(1.2.4) are, however, all about $r_\alpha(z,w)$ that we actually need in our argument.

We then consider $R_\alpha(z,w)$ defined by (1.1.51). We have to check the convergence of the series. To this end let us put

$$C(\eta;z,w) = \left\{\gamma \in \Gamma \;:\varrho(z,\gamma(w)) < \eta\right\} \quad (z = x+iy,\ w = u+iv \in \mathfrak{F}).$$

We have

$$C(\eta;z,w) = \bigcup_{\gamma\in\Gamma_\infty\backslash\Gamma} \left\{\gamma_1^n\gamma \;: n \in \mathbb{Z},\ \varrho(z,n+\gamma(w)) < \eta\right\},$$

where $\gamma_1(z) = z + 1$. The condition $\varrho(z, n + \gamma(w)) < \eta$ implies two inequalities,

$$(x - n - \operatorname{Re}\gamma(w))^2 < 4\eta y \operatorname{Im}\gamma(w), \quad (y - \operatorname{Im}\gamma(w))^2 < 4\eta y \operatorname{Im}\gamma(w).$$

From the first it follows that for each $\gamma \in \Gamma_\infty \backslash \Gamma$ there are at most $1 + 4(\eta y \operatorname{Im}\gamma(w))^{\frac{1}{2}}$ possible n's. From the second we have

$$y/q(\eta) < \operatorname{Im}\gamma(w) < q(\eta)y, \quad q(\eta) = 1 + 2\eta + 2(\eta(1+\eta))^{\frac{1}{2}}.$$

Thus we have

$$|C(\eta; z, w)| \ll \sum_M (1 + (\eta y M)^{\frac{1}{2}}) |\{\gamma \in \Gamma_\infty \backslash \Gamma \ : M \le \operatorname{Im}\gamma(w)\}|,$$

where M runs over the set of integral powers of 2 such that $y/(2q(\eta)) < M < 2q(\eta)y$. The last cardinality factor is not larger than the number of integral solutions l, h of the inequality $(lu + h)^2 + (lv)^2 \le v/M$; for cosets in $\Gamma_\infty \backslash \Gamma$ are parameterized by pairs of co-prime integers, as was indicated before deriving (1.1.5). Hence we have

$$|C(\eta; z, w)| \ll \sum_M (1 + (\eta y M)^{\frac{1}{2}})(1 + (Mv)^{-\frac{1}{2}})(1 + (v/M)^{\frac{1}{2}}).$$

This proves that we have, for any given $\eta_0 > 0$,

$$|C(\eta; z, w)| \ll \eta y + (\eta y v)^{\frac{1}{2}} \log(\eta y + 1) \quad (\eta > \eta_0), \tag{1.2.8}$$

where the implied constant depends only on η_0.

Now the third assertion in (1.2.4) and the last estimate yield that uniformly for $w \in \mathcal{F}$

$$R_\alpha(z, w) = \sum_{\substack{\gamma \in \Gamma \\ \varrho(z, \gamma(w)) < \eta_0}} r_\alpha(z, \gamma(w)) + O((\operatorname{Im} w)^{\frac{1}{2}}) \quad (z \in \mathcal{F}, \alpha > 1), \tag{1.2.9}$$

where the implied constant depends on α, η_0, and z. To see more closely the nature of this sum over γ we note that the condition $\varrho(z, w) < \eta_0$ is equivalent to the statement that w is in the Euclidean disk $\mathfrak{d}(z, \eta_0)$ of radius $2(\eta_0(1 + \eta_0))^{\frac{1}{2}} y$ with center at $(x, (1 + 2\eta_0)y)$. Let z be inside $\mathcal{F}$, and take η_0 so small that $\mathfrak{d}(z, \eta_0)$ is entirely inside $\mathcal{F}$. Then $w \in \mathcal{F} \backslash \mathfrak{d}(z, \eta_0)$ implies that the sum is empty; and $w \in \mathfrak{d}(z, \eta_0)$ implies that it has only one summand, i.e., the one corresponding to $\gamma = 1$. Hence we see, by the first assertion in (1.2.4), that uniformly for all $w \in \mathcal{F}$

$$R_\alpha(z, w) = -\frac{1}{2\pi} \log|z - w| + O((\operatorname{Im} w)^{\frac{1}{2}}) \quad (z \in \mathcal{F} \backslash \partial\mathcal{F}; \alpha > 1), \tag{1.2.10}$$

where the implied constant depends on α and z. On the other hand, if $z \in \partial\mathcal{F}$, then $R_\alpha(z,w)$, as a function of $w \in \mathcal{F}$, has logarithmic singularities at z and its equivalent boundary point as well. Note that the three points i, $\frac{1}{2}(\pm 1 + \sqrt{3}i)$ are special in this context, for each of them is equivalent to itself, i.e., a fixed point of certain non-trivial maps in Γ which are indicated at the end of the proof of Lemma 1.1. In any event the singularities of $R_\alpha(z,w)$ are of the logarithmic type only; and in particular the transformation (1.1.50) is well-defined throughout $B^\infty(\mathcal{F})$ whenever $\alpha > 1$. Naturally the whole situation becomes more transparent if we move to the manifold $\mathcal{F}$. It will turn out, however, that the above discussion is sufficient for our purpose (see the end of the proof of the following lemma).

We shall next show that $R_\alpha(z,w)$ induces the left-inverse of $\Delta+\alpha(\alpha-1)$ over $B^\infty(\mathcal{F})$:

Lemma 1.6 *When $\alpha > 1$ we have*

$$\mathcal{R}_\alpha(\Delta + \alpha(\alpha-1))f = f \qquad (1.2.11)$$

for any $f \in B^\infty(\mathcal{F})$.

Proof We prove first that the partial derivatives $\partial/\partial u$, $\partial/\partial v$, $(\partial/\partial u)^2$, $(\partial/\partial v)^2$, where $w = u + iv$, can be applied to the sum (1.1.51) termwise; and hence we have, by (1.2.3) and the invariance of Δ,

$$(\Delta + \alpha(\alpha-1))_w R_\alpha(z,w) = 0 \qquad \big(z, w \in \mathcal{H};\ z \not\equiv w \bmod \Gamma\big). \qquad (1.2.12)$$

Before confirming this, it should be noted that z can be assumed to be fixed, and that the condition $z \not\equiv w \bmod \Gamma$ implies that there exists a positive constant $c(z)$ such that $\varrho(z,\gamma(w)) \ge c(z)$ for all $\gamma \in \Gamma$ (see the argument prior to (1.2.10)).

We have

$$\begin{aligned}(\partial/\partial v) r_\alpha(z,\gamma(w)) &= (\partial/\partial v) r_\alpha(\gamma^{-1}(z), w) \\ &= p_\alpha'(\varrho(\gamma^{-1}(z), w))\varrho_v(\gamma^{-1}(z), w).\end{aligned}$$

Then, by (1.2.2) and (1.2.6), we have

$$\begin{aligned}(\partial/\partial v) r_\alpha(z,\gamma(w)) &\ll \varrho(z,\gamma(w))^{-\alpha} \frac{|v^2 - (\operatorname{Im}\gamma^{-1}(z))^2 - (u - \operatorname{Re}\gamma^{-1}(z))^2|}{v^2 \operatorname{Im}\gamma^{-1}(z)\varrho(\gamma^{-1}(z), w)} \\ &\ll \varrho(z,\gamma(w))^{-\alpha} \frac{v|v - \operatorname{Im}\gamma^{-1}(z)| + |w - \gamma^{-1}(z)|^2}{v|w - \gamma^{-1}(z)|^2} \\ &\ll \varrho(z,\gamma(w))^{-\alpha}(v^{-1} + |w - \gamma^{-1}(z)|^{-1}), \qquad (1.2.13)\end{aligned}$$

where the implied constant depends on $c(z)$ but not on w. This and (1.2.8) show that the sum

$$\sum_{\gamma\in\Gamma}(\partial/\partial v)r_\alpha(z,\gamma(w))$$

is locally uniformly convergent in the set $z \not\equiv w \bmod \Gamma$ for $\alpha > 1$; and it is equal to $(\partial/\partial v)R_\alpha(z,w)$. Similarly it can be shown that

$$\left.\begin{aligned}(\partial/\partial u)r_\alpha(z,\gamma(w)) &\ll \varrho(z,\gamma(w))^{-\alpha}|w-\gamma^{-1}(z)|^{-1},\\ (\partial/\partial u)^2 r_\alpha(z,\gamma(w)) &\ll \varrho(z,\gamma(w))^{-\alpha}|w-\gamma^{-1}(z)|^{-2},\\ (\partial/\partial v)^2 r_\alpha(z,\gamma(w)) &\ll \varrho(z,\gamma(w))^{-\alpha}(v^{-1}+|w-\gamma^{-1}(z)|^{-1})^2,\end{aligned}\right\} \tag{1.2.14}$$

which clearly ends the proof of (1.2.12).

We next turn to the proof of (1.2.11). We note that Δf is in $B^\infty(\mathcal{F})$, and thus the left side is well-defined. We have, for z lying inside $\mathcal{F}$,

$$\begin{aligned}&\mathcal{R}_\alpha(\Delta+\alpha(\alpha-1))f(z)\\ &\quad=\lim_{\delta\to 0}\int_{\mathcal{F}\setminus D_\delta(z)} R_\alpha(z,w)(\Delta+\alpha(\alpha-1))f(w)d\mu(w)\\ &\quad=\lim_{\delta\to 0}\int_{\mathcal{F}\setminus D_\delta(z)}\{R_\alpha(z,w)\Delta f(w)-f(w)\Delta_w R_\alpha(z,w)\}d\mu(w),\end{aligned}$$

where $D_\delta(z)$ is the Euclidean disk of radius δ with center at z, and we have used (1.2.10) and (1.2.12). We are going to apply Green's formula to the last integral. In this procedure we use the bounds for the relevant partial derivatives of $R_\alpha(z,w)$ which are readily obtainable from (1.2.8), (1.2.13), and (1.2.14); they are of polynomial order with respect to $\operatorname{Im} w$ as w tends to the cusp. Thus, on noting that $\partial\mathcal{F}$ contributes nothing, we have

$$\begin{aligned}&\mathcal{R}_\alpha(\Delta+\alpha(\alpha-1))f(z)\\ &\quad=\lim_{\delta\to 0}\int_{\partial D_\delta(z)}\{R_\alpha(z,w)(\partial/\partial n)_w f(w)-f(w)(\partial/\partial n)_w R_\alpha(z,w)\}|dw|\\ &\quad=-\lim_{\delta\to 0}\int_{\partial D_\delta(z)} f(w)(\partial/\partial n)_w R_\alpha(z,w)|dw|,\end{aligned} \tag{1.2.15}$$

where $\partial/\partial n$ is the ordinary outer-normal derivative. As before, it can be shown that we may apply $(\partial/\partial n)_w$ to (1.1.51) termwise, getting

$$(\partial/\partial n)_w R_\alpha(z,w)=\sum_{\gamma\in\Gamma}(\partial/\partial n)_w r_\alpha(z,\gamma(w)) \qquad \big(|w-z|=\delta\big).$$

If γ is not the identity map, we have obviously $\varrho(z,\gamma(w))\ge c(z)$, and thus, by (1.2.13) and (1.2.14),

$$(\partial/\partial n)_w r_\alpha(z,\gamma(w)) \ll \varrho(z,\gamma(w))^{-\alpha}$$

uniformly in δ. Hence we have, by the second assertion in (1.2.4) and (1.2.8),

$$(\partial/\partial n)_w R_\alpha(z,w) = -\frac{1}{2\pi\delta} + O(|\log\delta|) \qquad \big(|w-z| = \delta\big).$$

Inserting this into (1.2.15) we end the proof of (1.2.11) provided z is inside $\mathcal{F}$. The remaining case can be settled by the continuity argument. This ends the proof of the lemma.

1.3 Iterated kernel

Next we define the transformation $\mathcal{R}_{\alpha,n}$ in terms of the recursion formula

$$\mathcal{R}_{\alpha,n+1} = \big((n+1)(n+2\alpha)\big)^{-1}\big(\mathcal{R}_{\alpha,n} - \mathcal{R}_{\alpha+1,n}\big) \quad (n = 0,1,2,\ldots), \tag{1.3.1}$$

where $\mathcal{R}_{\alpha,0} = \mathcal{R}_\alpha$. We have

$$\mathcal{R}_{\alpha,N} = \sum_{\nu=0}^{N} \theta_{\nu,N}(\alpha)\mathcal{R}_{\alpha+\nu}, \tag{1.3.2}$$

where $\theta_{\nu,N}(\alpha)$ is a rational function of α, which is non-singular for $\alpha > 0$. It should be stressed that in this section we shall always assume

$$\alpha > 1, \qquad N \geq 1.$$

We have, for any $f \in B^\infty(\mathcal{F})$,

$$\mathcal{R}_{\alpha,N}\prod_{\nu=0}^{N}\big(\Delta + (\alpha+\nu-1)(\alpha+\nu)\big)f = f. \tag{1.3.3}$$

This can be proved by induction with respect to N: The initial step is (1.2.11). For arbitrary N we have

$$\begin{aligned}\mathcal{R}_{\alpha,N+1}&\prod_{\nu=0}^{N+1}\big(\Delta + (\alpha+\nu-1)(\alpha+\nu)\big)f = \big((N+1)(N+2\alpha)\big)^{-1}\\ &\times\Big\{\Big[\mathcal{R}_{\alpha,N}\prod_{\nu=0}^{N}\big(\Delta + (\alpha+\nu-1)(\alpha+\nu)\big)\Big]\big(\Delta + (\alpha+N)(\alpha+N+1)\big)f\\ &- \Big[\mathcal{R}_{\alpha+1,N}\prod_{\nu=0}^{N}\big(\Delta + (\alpha+\nu)(\alpha+\nu+1)\big)\Big]\big(\Delta + (\alpha-1)\alpha\big)f\Big\};\end{aligned}$$

and by the inductive hypothesis this is equal to

$$\big((N+1)(N+2\alpha)\big)^{-1}\Big\{(\Delta+(\alpha+N)(\alpha+N+1))f-(\Delta+(\alpha-1)\alpha)f\Big\}=f.$$

In particular we have

$$\mathcal{R}_{\alpha,N}f = q_{\alpha,N}(\lambda)f \qquad \big(f\in B^{\infty}(\mathcal{F}),\ \Delta f=\lambda f\big), \tag{1.3.4}$$

where

$$q_{\alpha,N}(\lambda)=\Big[\prod_{\nu=0}^{N}\big(\lambda+(\alpha+\nu)(\alpha+\nu-1)\big)\Big]^{-1}. \tag{1.3.5}$$

Thus $\mathcal{R}_{\alpha,N}$ is essentially diagonal over $\mathfrak{A}$. We have the decomposition

$$q_{\alpha,N}(\lambda)=\sum_{\nu=0}^{N}\theta_{\nu,N}(\alpha)\big(\lambda+(\alpha+\nu)(\alpha+\nu-1)\big)^{-1}. \tag{1.3.6}$$

To see the nature of the kernel function of $\mathcal{R}_{\alpha,N}$ we define $r_{\alpha,n}$ by

$$r_{\alpha,0}(z,w)=r_{\alpha}(z,w), \tag{1.3.7}$$

$$r_{\alpha,n}(z,w)-r_{\alpha+1,n}(z,w)=(n+1)(n+2\alpha)r_{\alpha,n+1}(z,w) \quad (n\ge 0), \tag{1.3.8}$$

where $r_{\alpha}(z,w)$ is as before. Obviously we have

$$r_{\alpha,N}(z,w)=\sum_{\nu=0}^{N}\theta_{\nu,N}(\alpha)r_{\alpha+\nu}(z,w). \tag{1.3.9}$$

From (1.1.51), (1.3.2), and (1.3.9) we get

$$\mathcal{R}_{\alpha,N}f(z)=\int_{\mathcal{F}}R_{\alpha,N}(z,w)f(w)d\mu(w),$$

where

$$R_{\alpha,N}(z,w)=\sum_{\gamma\in\Gamma}r_{\alpha,N}(z,\gamma(w)). \tag{1.3.10}$$

The definition (1.3.7)–(1.3.8) implies also

$$r_{\alpha,n}(z,w)=p_{\alpha,n}(\varrho(z,w)) \quad (n\ge 0), \tag{1.3.11}$$

where $\varrho(z,w)$ is as above, and

$$p_{\alpha,n}(\varrho)=\frac{\Gamma(\alpha)}{8\pi^2 i n!\Gamma(n+\alpha)}\int_{(-\frac{1}{4})}\frac{\Gamma(\xi+\alpha)\Gamma(\xi+\alpha+n)\Gamma(-\xi)}{\Gamma(\xi+2\alpha+n)}\varrho^{-\xi-\alpha}d\xi. \tag{1.3.12}$$

In fact this definition gives that $r_{\alpha,0}(z,w) = p_{\alpha,0}(\varrho(z,w))$ via (1.2.1), and that

$$p_{\alpha,n}(\varrho) - p_{\alpha+1,n}(\varrho) = (n+1)(n+2\alpha)p_{\alpha,n+1}(\varrho) \quad (n \geq 0),$$

which, in view of (1.3.8), proves (1.3.11).

For the sake of notational simplicity we now put

$$\left.\begin{aligned} R_\alpha^* = R_{\alpha,N}, \quad p_\alpha^* = p_{\alpha,N}, \quad q_\alpha^* = q_{\alpha,N}, \quad r_\alpha^* = r_{\alpha,N}, \quad \theta_\nu^* = \theta_{\nu,N}, \\ |(\partial z)^\nu F(z)| = |(\frac{\partial}{\partial x})^\nu F(z)| + |(\frac{\partial}{\partial y})^\nu F(z)|. \end{aligned}\right\} \quad (1.3.13)$$

Thus all expressions in this and the next sections possibly depend on the parameter N despite our neglecting to mention this.

Shifting the path in (1.3.12) with $n = N$ to $\operatorname{Re}\xi = -\alpha - \frac{1}{2}$ we see that $p_\alpha^*(\varrho)$ is continuous when ϱ tends to 0 from the right; and thus $r_\alpha^*(z,w)$ is continuous on the diagonal $z = w$. Then the discussion before and after (1.2.9) implies readily that R_α^* is continuous throughout $\mathfrak{F} \times \mathfrak{F}$. It is, however, unbounded, as has been suggested already. More precisely the asymptotic nature of R_α^* is as follows:

Lemma 1.7 *Let us assume*

$$\alpha > 1, \quad N \geq 4; \quad (1.3.14)$$

and put

$$R_\alpha^*(z,w) = \frac{1}{2\pi}\int_{-\infty}^{\infty} q_\alpha^*(t^2 + \tfrac{1}{4})(\operatorname{Im} z)^{\frac{1}{2}+it}(\operatorname{Im} w)^{\frac{1}{2}-it}dt + U_\alpha(z,w). \quad (1.3.15)$$

Then we have, uniformly for all $z = x+iy$, $w = u+iv$ in $\mathfrak{F}$,

$$|(\partial z)^\nu U_\alpha(z,w)| \ll y^{-\nu}\Big\{(vy)^{1-\alpha} + \Big(\frac{vy}{(v+y)^2}\Big)^\alpha\Big\} \quad (0 \leq \nu \leq 2). \quad (1.3.16)$$

Proof We shall exploit the decomposition of $R_\alpha(z,w)$, which is implicit in the argument leading to (1.2.9). Thus we put

$$R_\alpha^*(z,w) = \sum_{\gamma\in\Gamma_\infty\backslash\Gamma} S_\alpha(z,\gamma(w)),$$

where

$$S_\alpha(z,w) = \sum_{n=-\infty}^{\infty} r_\alpha^*(z,n+w).$$

By partial summation we have

$$S_\alpha(z,w) = \int_{-\infty}^{\infty} r_\alpha^*(z,t+w)dt + \int_{-\infty}^{\infty} \xi(t)(\partial/\partial t)r_\alpha^*(z,t+w)dt$$
$$= \{S_\alpha^{(0)} + S_\alpha^{(1)}\}(z,w), \tag{1.3.17}$$

say, where $\xi(t) = t-[t]-\frac{1}{2}$. That these integrals are absolutely convergent is clear from the discussion below. We then have the decomposition

$$R_\alpha^*(z,w) = S_\alpha^{(0)}(z,w) + T_\alpha^{(0)}(z,w) + T_\alpha^{(1)}(z,w), \tag{1.3.18}$$

where

$$T_\alpha^{(0)}(z,w) = \sum_{\substack{\gamma\in\Gamma_\infty\backslash\Gamma \\ \gamma\notin\Gamma_\infty}} S_\alpha^{(0)}(z,\gamma(w))$$

with a common abuse of notation, and

$$T_\alpha^{(1)}(z,w) = \sum_{\gamma\in\Gamma_\infty\backslash\Gamma} S_\alpha^{(1)}(z,\gamma(w)).$$

In order to evaluate $S_\alpha^{(0)}(z,w)$ explicitly we shall first prove that

$$\int_{\mathcal{H}} r_\alpha(z,w)(\operatorname{Im} w)^s d\mu(w) = (s(1-s)+\alpha(\alpha-1))^{-1}(\operatorname{Im} z)^s \tag{1.3.19}$$

provided

$$\alpha > \operatorname{Re} s > 1-\alpha. \tag{1.3.20}$$

This could be proved via (1.1.19), but we shall take an alternative way to indicate a variety of available methods: The first and the last assertions in (1.2.4) give

$$\int_{\mathcal{H}} |r_\alpha(z,w)(\operatorname{Im} w)^s| d\mu(w)$$
$$\ll 1 + \iint_{\mathcal{H},\,|w-z|\ge\frac{1}{2}y} \frac{v^{\alpha+\operatorname{Re} s-2}}{((u-x)^2+(v-y)^2)^\alpha} du dv, \tag{1.3.21}$$

which is obviously finite if (1.3.20) holds. Thus we have, given (1.3.20),

$$\int_{\mathcal{H}} r_\alpha(z,w)(\operatorname{Im} w)^s d\mu(w) = \lim_{\delta\to+0} \int_{|v-y|>\delta} r_\alpha(z,w)(\operatorname{Im} w)^s d\mu(w). \tag{1.3.22}$$

Into this we insert (1.2.2). The resulting triple integral is absolutely convergent, and the right side of (1.3.22) can be written as

$$\frac{1}{8\pi^2 i}\lim_{\delta\to+0}\left\{\int_0^{y-\delta}+\int_{y+\delta}^{\infty}\right\}v^{s-2}\int_{(-\frac{1}{4})}\frac{\Gamma(\xi+\alpha)^2\Gamma(-\xi)}{\Gamma(\xi+2\alpha)}$$
$$\times\int_{-\infty}^{\infty}\left(\frac{4vy}{(u-x)^2+(v-y)^2}\right)^{\xi+\alpha}du\,d\xi\,dv. \tag{1.3.23}$$

The double integral over u, ξ is equal to

$$(4\pi vy)^{\frac{1}{2}}\int_{(-\frac{1}{4})}\frac{\Gamma(\xi+\alpha)\Gamma(\xi+\alpha-\frac{1}{2})\Gamma(-\xi)}{\Gamma(\xi+2\alpha)}\left(\frac{4vy}{(v-y)^2}\right)^{\xi+\alpha-\frac{1}{2}}d\xi$$
$$=2\pi i(4\pi vy)^{\frac{1}{2}}\left(\frac{4vy}{(v-y)^2}\right)^{\alpha-\frac{1}{2}}\frac{\Gamma(\alpha)\Gamma(\alpha-\frac{1}{2})}{\Gamma(2\alpha)}F\left(\alpha,\alpha-\tfrac{1}{2};2\alpha;-\frac{4vy}{(v-y)^2}\right). \tag{1.3.24}$$

The duplication formula for the Γ-function and the well-known identity

$$F(\alpha,\alpha-\tfrac{1}{2};2\alpha;\tau)=(\tfrac{1}{2}(1+\sqrt{1-\tau}))^{1-2\alpha}\qquad(|\arg(1-\tau)|<\pi)$$

simplify the right side of (1.3.24) to

$$8\pi^2 i(2\alpha-1)^{-1}\left(\tfrac{1}{2}(v+y+|v-y|)\right)^{1-2\alpha}(vy)^{\alpha}.$$

Thus we have, by (1.3.22) and (1.3.23),

$$\int_{\mathcal{H}}r_\alpha(z,w)(\operatorname{Im}w)^s d\mu(w)$$
$$=(2\alpha-1)^{-1}y^{\alpha}\int_0^{\infty}\left(\tfrac{1}{2}(v+y+|v-y|)\right)^{1-2\alpha}v^{s+\alpha-2}dv,$$

which gives rise to (1.3.19).

Then we consider the Mellin transform of $S_\alpha^{(0)}(iy,iv)v^{-1}$ with respect to the variable v; here it is to be observed that we have $S_\alpha^{(0)}(z,w)=S_\alpha^{(0)}(iy,iv)$ with v,y as above. We have, given (1.3.20),

$$\int_0^{\infty}S_\alpha^{(0)}(z,w)v^{s-2}dv=\int_{\mathcal{H}}r_\alpha^*(z,w)(\operatorname{Im}w)^s d\mu(w)$$
$$=\sum_{\nu=0}^{N}\theta_\nu^*(\alpha)\int_{\mathcal{H}}r_{\alpha+\nu}(z,w)(\operatorname{Im}w)^s d\mu(w),$$

where the integrals are absolutely convergent because of (1.3.21), and the relation (1.3.9) has been used. By virtue of (1.3.19) we get, via (1.3.6),

$$\int_0^{\infty}S_\alpha^{(0)}(z,w)v^{s-2}dv=q_\alpha^*(s(1-s))(\operatorname{Im}z)^s$$

provided (1.3.20) holds. In this the integrand is continuous with respect to v and the right side is sufficiently smooth with respect to s. Hence Mellin's inversion formula yields

$$S_\alpha^{(0)}(z,w) = \frac{1}{2\pi}\int_{-\infty}^{\infty} q_\alpha^*(t^2+\tfrac{1}{4})(\operatorname{Im} z)^{\frac{1}{2}+it}(\operatorname{Im} w)^{\frac{1}{2}-it}dt. \tag{1.3.25}$$

Shifting the path to $\operatorname{Im} t = \pm\infty$ and invoking the formula (1.3.6), we get

$$S_\alpha^{(0)}(z,w) = \sum_{\nu=0}^{N} \frac{\theta_\nu^*(\alpha)}{2\alpha+2\nu-1} \times \begin{cases} (\operatorname{Im} w)^{\alpha+\nu}(\operatorname{Im} z)^{1-\alpha-\nu} & \text{if } \operatorname{Im} w \le \operatorname{Im} z, \\ (\operatorname{Im} z)^{\alpha+\nu}(\operatorname{Im} w)^{1-\alpha-\nu} & \text{if } \operatorname{Im} z \le \operatorname{Im} w. \end{cases} \tag{1.3.26}$$

This implies that providing $w,\, z \in \mathcal{F}$, $\operatorname{Im} z \ge 1$, we have

$$T_\alpha^{(0)}(z,w) = \sum_{\nu=0}^{N} \frac{\theta_\nu^*(\alpha)}{2\alpha+2\nu-1}(\operatorname{Im} z)^{1-\alpha-\nu}\{E(w,\alpha+\nu) - (\operatorname{Im} w)^{\alpha+\nu}\}, \tag{1.3.27}$$

where E is the Eisenstein series; and we have used the fact that $\operatorname{Im}\gamma(w) \le 1$ for $\gamma \notin \Gamma_\infty$. Then we find that uniformly for $w, z \in \mathcal{F}$

$$\left(\frac{\partial}{\partial x}\right) T_\alpha^{(0)}(z,w) = 0, \quad \left(\frac{\partial}{\partial y}\right)^\nu T_\alpha^{(0)}(z,w) \ll y^{-\nu}(vy)^{1-\alpha} \quad (\nu \ge 0). \tag{1.3.28}$$

If $\operatorname{Im} z \ge 1$, this follows readily from (1.1.9) and (1.3.27). Otherwise we may of course assume that $\operatorname{Im} w \ge 2$, say; and we can again use (1.3.27). This ends the treatment of $T_\alpha^{(0)}$.

We turn to $T_\alpha^{(1)}$. We have to estimate the derivatives of $S_\alpha^{(1)}(z,w)$. We are going to prove that providing (1.3.14) holds we have, uniformly for $w,\, z \in \mathcal{H}$,

$$|(\partial z)^\nu S_\alpha^{(1)}(z,w)| \ll y^{-\nu}\left(\frac{vy}{(v+y)^2}\right)^\alpha \quad (0 \le \nu \le 2). \tag{1.3.29}$$

The confirmation of this estimate will finish the proof of the lemma. For (1.3.29) yields that for $w, z \in \mathcal{F}$

$$\begin{aligned} |(\partial z)^\nu T_\alpha^{(1)}(z,w)| &\ll y^{-\nu}\left(\frac{vy}{(v+y)^2}\right)^\alpha + y^{-\nu-\alpha}\{E(w,\alpha) - (\operatorname{Im} w)^\alpha\} \\ &\ll y^{-\nu}\left(\frac{vy}{(v+y)^2}\right)^\alpha + y^{-\nu-\alpha}v^{1-\alpha}, \end{aligned}$$

where we have used (1.1.9). Collecting this and (1.3.18), (1.3.25), (1.3.28), we obtain (1.3.16). As to the proof of (1.3.29), we shall deal with only $(\partial/\partial y)^2 S_\alpha^{(1)}(z,w)$ in detail, since the estimate of other parts is simpler. To this end we note that the differentiation of $S_\alpha^{(1)}(z,w)$ can be performed inside the defining integral. For the resulting integral converges uniformly, as is clear from our estimates below. Thus we have, by (1.3.11),

$$\Big(\frac{\partial}{\partial y}\Big)^2 S_\alpha^{(1)}(z,w) = \int_{-\infty}^{\infty} \xi(t)\big[(p_\alpha^*)'''(\varrho)\varrho_y^2\varrho_u + (p_\alpha^*)''(\varrho)\varrho_{yy}\varrho_u + 2(p_\alpha^*)''(\varrho)\varrho_y\varrho_{uy} + (p_\alpha^*)'(\varrho)\varrho_{uyy}\big]dt, \quad (1.3.30)$$

where $\varrho = \varrho(z, t+w)$. The first term of the integrand contributes, by (1.2.6),

$$\ll \int_0^\infty \Big|(p_\alpha^*)'''\Big(\frac{t^2+(v-y)^2}{4vy}\Big)\Big|\Big(\frac{t^2+v^2-y^2}{vy^2}\Big)^2\frac{t}{vy}dt$$
$$\ll y^{-2}\int_0^\infty |(p_\alpha^*)'''(t^2+\varrho(iy,iv))|\big(t^5+(v/y-y/v)^2t\big)dt. \quad (1.3.31)$$

We now use (1.3.12) with the present supposition $N \ge 4$; we see readily that it gives

$$(p_\alpha^*)'''(\tau) \ll \min\{1, \tau^{-\alpha-3}\} \quad (\tau > 0).$$

This implies that (1.3.31) is not larger than the right side of (1.3.29) with $\nu = 2$. Other parts of (1.3.30) can be estimated in just the same way. This ends the verification of (1.3.29); and we finish the proof of Lemma 1.7.

Next we shall study the function

$$C_\alpha(z,w) = \frac{1}{4\pi}\int_{-\infty}^{\infty} q_\alpha^*(t^2+\tfrac14)E(z,\tfrac12+it)E(w,\tfrac12-it)dt + \frac{3}{\pi}q_\alpha^*(0). \quad (1.3.32)$$

The reason for the presence of the last term will soon become obvious. Combining the results on C_α to be proved below with the assertions of the preceding lemma we shall obtain the following compactification of the transformation $\mathcal{R}_\alpha^*$:

Lemma 1.8 *Let us assume* (1.3.14), *and put*

$$R_\alpha^*(z,w) = \tilde{R}_\alpha(z,w) + C_\alpha(z,w). \quad (1.3.33)$$

Then we have

$$\int_{\mathcal{F}} \tilde{R}_\alpha(z,w)f(w)d\mu(w) = q_\alpha^*(\lambda)f(z) \quad (1.3.34)$$

provided $\Delta f = \lambda f$, $f \in B^\infty(\mathcal{F})$. *Also we have, uniformly for all* $z = x+iy$, $w = u+iv$ *in* $\mathcal{F}$,

$$|(\partial z)^\nu \tilde{R}_\alpha(z,w)| \ll y^{-\nu}\Big\{(vy)^{1-\alpha}+(vy)^{2-N}+\Big(\frac{vy}{(v+y)^2}\Big)^\alpha\Big\} \quad (0 \le \nu \le 2). \quad (1.3.35)$$

In particular, $\tilde{R}_\alpha(z,w)$ *is a bounded continuous function on* $\mathcal{F}\times\mathcal{F}$.

Proof The absolute convergence of the integral in (1.3.32) will become clear in the course of discussion. We write the expansion (1.1.9) as

$$E(z,s) = y^s + \varphi_\Gamma(s)y^{1-s} + \tilde{E}(z,s), \qquad (1.3.36)$$

and observe that $\varphi_\Gamma(s)\varphi_\Gamma(1-s) = 1$ because of (1.1.13) as well as that (1.1.11) gives

$$\tilde{E}(z,s) = \varphi_\Gamma(s)\tilde{E}(z,1-s).$$

Then we have

$$C_\alpha(z,w) = \frac{1}{2\pi}\int_{-\infty}^{\infty} q_\alpha^*(t^2+\tfrac{1}{4})y^{\frac{1}{2}+it}v^{\frac{1}{2}-it}dt$$
$$+ C_\alpha^{(0)}(z,w) + C_\alpha^{(1)}(z,w) + C_\alpha^{(1)}(w,z) + C_\alpha^{(2)}(z,w), \qquad (1.3.37)$$

where

$$C_\alpha^{(0)}(z,w) = \frac{1}{2\pi}\int_{-\infty}^{\infty} q_\alpha^*(t^2+\tfrac{1}{4})(vy)^{\frac{1}{2}-it}\varphi_\Gamma(\tfrac{1}{2}+it)dt + \frac{3}{\pi}q_\alpha^*(0),$$

$$C_\alpha^{(1)}(z,w) = \frac{1}{2\pi}\int_{-\infty}^{\infty} q_\alpha^*(t^2+\tfrac{1}{4})v^{\frac{1}{2}-it}\tilde{E}(z,\tfrac{1}{2}+it)dt, \qquad (1.3.38)$$

$$C_\alpha^{(2)}(z,w) = \frac{1}{4\pi}\int_{-\infty}^{\infty} q_\alpha^*(t^2+\tfrac{1}{4})\tilde{E}(z,\tfrac{1}{2}+it)\tilde{E}(w,\tfrac{1}{2}-it)dt. \qquad (1.3.39)$$

We have, by (1.3.15), (1.3.33), and (1.3.37),

$$\tilde{R}_\alpha(z,w) = U_\alpha(z,w) - C_\alpha^{(0)}(z,w) - C_\alpha^{(1)}(z,w) - C_\alpha^{(1)}(w,z) - C_\alpha^{(2)}(z,w). \quad (1.3.40)$$

We are going to estimate the partial derivatives of $C_\alpha^{(j)}$. In doing this we may assume $vy \ge 1$ without loss of generality. In fact the crucial estimates (1.3.41), (1.3.49)–(1.3.51) below are trivial if $vy < 1$.

The case $j = 0$ is easy: In the integral for $C_\alpha^{(0)}$ we shift the path to $\mathrm{Im}\, t = -\infty$, encountering simple poles at $t = -\frac{1}{2}i$ and $-i(\alpha + \nu - \frac{1}{2})$ $(\nu = 0, 1, \ldots, N)$. We get

$$C_\alpha^{(0)}(z,w) = \sum_{\nu=0}^{N} \tau_{\nu,N}(\alpha)(vy)^{1-\alpha-\nu}$$

with certain $\tau_{\nu,N}(\alpha)$; note that the residue of the pole at $t = -\frac{1}{2}i$ cancels the last term in (1.3.32) out. This identity implies immediately that uniformly for w, $z \in \mathcal{F}$

$$|(\partial z)^\nu C_\alpha^{(0)}(z,w)| \ll y^{-\nu}(vy)^{1-\alpha}. \qquad (1.3.41)$$

Before dealing with the cases $j = 1, 2$ we remark that, as the estimations below will ensure, the relevant differentiation with respect to x, y

can be performed inside the integrals in (1.3.38)–(1.3.39); and thus the derivatives of $\tilde{E}(z,s)$ come into consideration. But we have, by (1.1.12),

$$\Delta\tilde{E}(z,s) = s(1-s)\tilde{E}(z,s). \tag{1.3.42}$$

That is, we have

$$\left(\frac{\partial}{\partial y}\right)^2 \tilde{E}(z, \tfrac{1}{2}+it) = -\left[y^{-2}(t^2+\tfrac{1}{4}) + \left(\frac{\partial}{\partial x}\right)^2\right]\tilde{E}(z,\tfrac{1}{2}+it); \tag{1.3.43}$$

and, as the right side decays exponentially with respect to y, we have

$$\frac{\partial}{\partial y}\tilde{E}(z,\tfrac{1}{2}+it) = -\int_y^\infty \left(\frac{\partial}{\partial y}\right)^2 \tilde{E}(z,\tfrac{1}{2}+it)dy. \tag{1.3.44}$$

These imply that it is sufficient to consider the estimation of $(\partial/\partial x)^\nu C_\alpha^{(j)}(z,w)$ or rather that of $(\partial/\partial x)^\nu \tilde{E}(z,\frac{1}{2}+it)$.

To this end we show an estimate for the Bessel function $K_{it}(y)$ with complex t : We put $\nu = it$ in (1.1.16), and integrate by parts, getting

$$K_{it}(y) = -i2^{it}\frac{\Gamma(\frac{3}{2}+it)}{\pi^{\frac{1}{2}}y^{1+it}}\tilde{K}_{it}(y), \tag{1.3.45}$$

where

$$\tilde{K}_{it}(y) = \int_{-\infty}^{\infty}\frac{\xi e^{iy\xi}}{(1+\xi^2)^{\frac{3}{2}+it}}d\xi. \tag{1.3.46}$$

Shifting the path to $\operatorname{Im}\xi = 1/(1+|t|)$, we see that

$$\tilde{K}_{it}(y) \ll \exp(-y/(1+|t|)) \qquad (y>0,\ \operatorname{Im} t \le \tfrac{1}{4}), \tag{1.3.47}$$

where the implied constant is absolute.

These yield that $\tilde{E}(z,\frac{1}{2}+it)$ is regular for $\operatorname{Im} t \le 0$; and in this region we have

$$\begin{aligned}
&\left(\frac{\partial}{\partial x}\right)^\nu \tilde{E}(z,\tfrac{1}{2}+it) \\
&= \frac{(\frac{1}{2}+it)(2\pi i)^\nu}{\pi i\zeta(1+2it)y^{\frac{1}{2}+it}}\sum_{n\ne 0} n^\nu |n|^{-1}\sigma_{-2it}(|n|)\tilde{K}_{it}(2\pi|n|y)e(nx) \\
&\ll \frac{(1+|t|)y^{-\frac{1}{2}+\operatorname{Im} t}}{|\zeta(1+2it)|}\sum_{n=1}^\infty n^\nu \exp(-2\pi yn/(1+|t|)) \\
&\ll \left(\frac{1+|t|}{y}\right)^{\nu+2}\frac{y^{\frac{1}{2}+\operatorname{Im} t}}{|\zeta(1+2it)|}\exp(-\pi y/(1+|t|)),
\end{aligned} \tag{1.3.48}$$

where the implied constants are absolute. Having this, we may return to (1.3.38). We differentiate inside the integral and shift the path to

$\operatorname{Im} t = -\infty$; note that we have $q_\alpha^*(t^2 + \frac{1}{4}) = O((1+|t|)^{-2N-2})$, $N \geq 4$, and $0 \leq \nu \leq 2$. Computing the residues arising from the poles of $q_\alpha^*(t^2 + \frac{1}{4})$ we get

$$\left(\frac{\partial}{\partial x}\right)^\nu C_\alpha^{(1)}(z,w) \ll v^{1-\alpha} \exp(-y/(\alpha+N)).$$

In view of (1.3.43)–(1.3.44) the same bound holds for $(\partial/\partial y)^\nu C_\alpha^{(1)}(z,w)$ as well. Thus we have

$$|(\partial z)^\nu C_\alpha^{(1)}(z,w)| \ll v^{1-\alpha} \exp(-y/(\alpha+N)). \tag{1.3.49}$$

Similarly we have

$$|(\partial z)^\nu C_\alpha^{(1)}(w,z)| \ll y^{1-\alpha-\nu} \exp(-v/(\alpha+N)). \tag{1.3.50}$$

As to the case $j = 2$ we use (1.3.48) with real t, getting readily

$$\begin{aligned} |(\partial z)^\nu C_\alpha^{(2)}(z,w)| &\ll y^{-\nu}(vy)^{-\frac{3}{2}} \int_{-\infty}^{\infty} \frac{\exp(-\pi(v+y)/(1+|t|))}{|\zeta(1+2it)|^2(1+|t|)^{2N-\nu-2}} dt \\ &\ll y^{-\nu}(vy)^{1-N+\frac{1}{2}\nu}. \end{aligned} \tag{1.3.51}$$

Combining (1.3.16), (1.3.40), (1.3.41), and (1.3.49)–(1.3.51) we end the proof of (1.3.35).

Next, in view of (1.3.4) the assertion (1.3.34) is equivalent to the claim

$$\int_{\mathcal{F}} C_\alpha(z,w) f(w) d\mu(w) = 0 \qquad \left(f \in B^\infty(\mathcal{F}),\ \Delta f = \lambda f\right). \tag{1.3.52}$$

We shall prove this: We insert the definition (1.3.32) into the left side. The resulting expression is a triple integral, for we have $\langle f, 1 \rangle = 0$, as we remarked already. It is absolutely convergent because of (1.3.48) with $\nu = 0$. Thus we have, after an exchange of the order of integration,

$$\int_{\mathcal{F}} C_\alpha(z,w) f(w) d\mu(w) = \lim_{\delta \to +0} \int_{\substack{-\infty \\ |t^2-\kappa^2|>\delta}}^{\infty} q_\alpha^*(t^2 + \tfrac{1}{4}) E(z, \tfrac{1}{2} + it) \mathcal{E}(t,f) dt, \tag{1.3.53}$$

where $\lambda = \kappa^2 + \frac{1}{4}$ with a positive κ, and $\mathcal{E}(t,f)$ is defined by (1.1.46); note that in the present situation the relevant limit in mean reduces to the ordinary one. Hence it is enough to show that

$$\mathcal{E}(t,f) = 0 \qquad \left(t^2 \neq \kappa^2\right). \tag{1.3.54}$$

To this end we invoke the relation

$$(t^2 - \kappa^2)\mathcal{E}(t,f) = \int_{\mathcal{F}} \left[f(w) \Delta E(w, \tfrac{1}{2} - it) - E(w, \tfrac{1}{2} - it) \Delta f(w)\right] d\mu(w),$$

which depends on (1.1.12). By Green's formula we have

$$\begin{aligned}(t^2-\kappa^2)\mathcal{E}(t,f)\\ &= -\int_{\partial\mathfrak{F}}[f(w)v\frac{\partial}{\partial n}E(w,\tfrac12-it)-E(w,\tfrac12-it)v\frac{\partial}{\partial n}f(w)]\frac{|dw|}{v}\\ &=0,\end{aligned}$$

where the reasoning is precisely the same as for (1.1.31). This ends the proof of the lemma.

1.4 Spectral resolution

The aim of this section is to finish the proof of Theorem 1.1. All clues are given in the last lemma. We begin by confirming the first assertion of the theorem:

Lemma 1.9 *The set of eigenvalues of* Δ *acting over* $B^\infty(\mathfrak{F})$ *is discrete. Hence we have*

$$\mathfrak{A} = \text{the subspace spanned by the orthonormal system } \{\psi_j\ :\ j\geq 1\},$$

where the ψ_j*'s are the Maass waves introduced in the first section.*

Proof This is a typical instance of the applications of Bessel's inequality: Let $\{f_\lambda\ :\ \lambda\in\Lambda\}$ be an arbitrary finite orthonormal system in $B^\infty(\mathfrak{F})$ such that $\Delta f_\lambda=\lambda f_\lambda$. On the assumption (1.3.14) we obviously have

$$0\leq\int_{\mathfrak{F}}|\tilde{R}_\alpha(z,w)-\sum_{\lambda\in\Lambda}q_\alpha^*(\lambda)f_\lambda(z)\overline{f_\lambda(w)}\,|^2d\mu(w).$$

Expanding out the square and invoking the assertion (1.3.34) we get

$$\sum_{\lambda\in\Lambda}q_\alpha^*(\lambda)^2|f_\lambda(z)|^2\leq\int_{\mathfrak{F}}|\tilde{R}_\alpha(z,w)|^2d\mu(w). \tag{1.4.1}$$

The right side is bounded uniformly for all $z\in\mathfrak{F}$ because of (1.3.35), $v=0$. Integrating this we obtain

$$\sum_{\lambda\in\Lambda}q_\alpha^*(\lambda)^2\leq\int_{\mathfrak{F}\times\mathfrak{F}}|\tilde{R}_\alpha(z,w)|^2d\mu(w)d\mu(z)<\infty, \tag{1.4.2}$$

which gives rise to the assertion in the lemma.

What we are going to establish next is that the transformation induced by the kernel $\tilde{R}_\alpha(z,w)$ is essentially the projection to the subspace $\mathfrak{A}$:

Lemma 1.10 *Providing*

$$\alpha \geq \tfrac{3}{2}, \quad N \geq 20, \tag{1.4.3}$$

we have that

$$\tilde{R}_\alpha(z,w) = \sum_{j=1}^{\infty} q_\alpha^*(\lambda_j)\psi_j(z)\overline{\psi_j(w)}. \tag{1.4.4}$$

That is,

$$R_\alpha^*(z,w) = \sum_{j=0}^{\infty} q_\alpha^*(\lambda_j)\psi_j(z)\overline{\psi_j(w)} + \frac{1}{4\pi}\int_{-\infty}^{\infty} q_\alpha^*(t^2+\tfrac{1}{4})E(z,\tfrac{1}{2}+it)E(w,\tfrac{1}{2}-it)dt, \tag{1.4.5}$$

where $\lambda_0 = 0$, $\psi_0 \equiv (3/\pi)^{\frac{1}{2}}$; and the convergence is absolute and locally uniform for all w, $z \in \mathcal{F}$.

Proof We note first that if $\alpha > 1$ and $N \geq 9$ then the sum in (1.4.4) converges absolutely and uniformly throughout $\mathcal{F} \times \mathcal{F}$. For we have (1.4.1), given (1.3.14), and also $q_\alpha^*(\lambda) \approx \lambda^{-N-1}$. Thus we may put

$$\hat{R}_\alpha(z,w) = \tilde{R}_\alpha(z,w) - D_\alpha(z,w), \tag{1.4.6}$$

where $D_\alpha(z,w)$ is the right side of (1.4.4). Then we shall show that providing (1.4.3) holds we have, uniformly for all $w = u+vi$, $z = x+yi \in \mathcal{F}$,

$$|(\partial z)^\nu \hat{R}_\alpha(z,w)| \ll y^{-\nu}\left\{(vy)^{-\frac{1}{2}} + \left(\frac{vy}{(v+y)^2}\right)^\alpha\right\} \quad (0 \leq \nu \leq 2). \tag{1.4.7}$$

To prove this it is sufficient to have the crude bound

$$|(\partial z)^\nu \psi_j(z)| \ll y^{-\nu-\frac{1}{2}}\lambda_j^5 \quad (0 \leq \nu \leq 2,\ z \in \mathcal{F}). \tag{1.4.8}$$

In fact this estimate implies that if $\alpha > 1$ and $N \geq 20$ then

$$|(\partial z)^\nu D_\alpha(z,w)| \ll y^{-\nu}(vy)^{-\frac{1}{2}} \sum_{j=1}^{\infty} \lambda_j^{-10} \ll y^{-\nu}(vy)^{-\frac{1}{2}}, \tag{1.4.9}$$

where we have used (1.4.2) with $N = 4$. This and (1.3.35) yield (1.4.7) immediately. On the other hand, in order to prove (1.4.8) we note that the relation (1.3.34) gives, on the assumption (1.3.14),

$$q_\alpha^*(\lambda_j)|(\partial z)^\nu \psi_j(z)| \leq \int_{\mathcal{F}} |(\partial z)^\nu \tilde{R}_\alpha(z,w)||\psi_j(w)|d\mu(w). \tag{1.4.10}$$

Thus we have, by (1.3.35) with $N = 4$,

$$
\begin{aligned}
|(\partial z)^\nu \psi_j(z)|^2 &\ll \lambda_j^{10} \int_{\mathcal{F}} |(\partial z)^\nu \tilde{R}_\alpha(z,w)|^2 d\mu(w) \\
&\ll \lambda_j^{10} y^{-2\nu} \int_{\frac{1}{2}}^{\infty} \Big[(vy)^{2(1-\alpha)} + (vy)^{-4} + \Big(\frac{vy}{(v+y)^2}\Big)^{2\alpha}\Big] v^{-2} dv \\
&\ll \lambda_j^{10} y^{-2\nu}\big(y^{2(1-\alpha)} + y^{-1}\big). \qquad (1.4.11)
\end{aligned}
$$

This gives (1.4.8); and we end the proof of (1.4.7).

Continuing the argument, we observe that

$$\hat{R}_\alpha(z,w) = \overline{\hat{R}_\alpha(w,z)}$$

by the the chain of expressions (1.2.1), (1.3.9), (1.3.10), (1.3.13), (1.3.32), (1.3.33), and (1.4.6). Also the estimate (1.4.7) implies, in particular, that $\hat{R}_\alpha(z,w)$ is bounded and continuous; in fact it tends to 0, e.g., as $z \to i\infty$ while w remains bounded. Thus we may appeal to the Hilbert–Schmidt theory on integral equations with bounded continuous symmetric kernels: We assume that (1.4.4) does not hold. Then there should exist a real $\lambda \neq 0$ for which the integral equation on the manifold $\mathcal{F}$,

$$\int_{\mathcal{F}} \hat{R}_\alpha(z,w) f(w) d\mu(w) = \lambda f(z), \qquad (1.4.12)$$

has a non-trivial solution f in $L^2(\mathcal{F}, d\mu)$. We denote by L_λ the set of all such solutions. Obviously L_λ is a finite dimensional subspace of $L^2(\mathcal{F}, d\mu)$. We then observe that $L_\lambda \subseteq C^2(\mathcal{F})$. Or more precisely, (1.4.12) implies that

$$|(\partial z)^\nu f(z)| \ll y^{-\nu-\frac{1}{2}} \quad (0 \le \nu \le 2,\ z \in \mathcal{F}). \qquad (1.4.13)$$

This follows from (1.4.7); and the argument is analogous to (1.4.10)–(1.4.11). Having this, we can show that Δ induces a symmetric linear transformation in L_λ; i.e., we are going to show that

$$\Delta L_\lambda \subseteq L_\lambda \qquad (1.4.14)$$

and

$$\langle \Delta f, g\rangle = \langle f, \Delta g\rangle \qquad (f, g \in L_\lambda). \qquad (1.4.15)$$

To prove (1.4.14) we note that the assertion (1.4.7) allows us to apply Δ to both sides of (1.4.12), and to exchange the order of Δ and the integration on the left side. Thus we have

$$\int_{\mathcal{F}} \Delta_z \hat{R}_\alpha(z,w) f(w) d\mu(w) = \lambda \Delta f(z). \qquad (1.4.16)$$

But we have

$$\Delta_z \hat{R}_\alpha(z,w) = \Delta_w \hat{R}_\alpha(z,w).$$

To verify this we check again the chain of expressions mentioned above; then we see that it is sufficient to have the same identity for each of $D_\alpha(z,w)$, $C_\alpha(z,w)$, and $R_\alpha^*(z,w)$, in place of $\hat{R}_\alpha(z,w)$. The case of $D_\alpha(z,w)$ is confirmed by applying Δ_z to the sum in (1.4.4) termwise, which is legitimate in view of (1.4.8). The case of $C_\alpha(z,w)$ is similar. Further, the case of $R_\alpha^*(z,w)$ is readily reduced to (1.2.5) and the invariance of Δ. Hence we have, instead of (1.4.16),

$$\int_{\mathfrak{F}} \Delta_w \hat{R}_\alpha(z,w) f(w) d\mu(w) = \lambda \Delta f(z);$$

and Green's formula gives

$$\int_{\mathfrak{F}} \hat{R}_\alpha(z,w) \Delta f(w) d\mu(w) = \lambda \Delta f(z).$$

This ends the proof of (1.4.14), since $\Delta f \in L^2(\mathfrak{F}, d\mu)$ is obvious because of (1.4.13). On the other hand the identity (1.4.15) can be proved as well by Green's formula, and the details can be omitted.

Hence there should be a non-trivial $f \in L_\lambda$ such that

$$\Delta f = \lambda' f$$

with a certain real λ'. We have $\lambda' \neq 0$. For otherwise f would be a constant, as was shown in the proof of Lemma 1.4; but then the estimate (1.4.13) with $v = 0$ would imply $f \equiv 0$. Thus we may now appeal to Lemma 1.4, which yields under the present situation that $f \in \mathfrak{A}$. In particular we have, by (1.3.34),

$$\int_{\mathfrak{F}} \tilde{R}_\alpha(z,w) f(w) d\mu(w) = q_\alpha^*(\lambda') f(z).$$

Also, expressing f as a linear combination of ψ_j's such that $\lambda_j = \lambda'$ while noting (1.1.42) we see readily that

$$\int_{\mathfrak{F}} D_\alpha(z,w) f(w) d\mu(w) = q_\alpha^*(\lambda') f(z).$$

That is, we have, by (1.4.6),

$$\int_{\mathfrak{F}} \hat{R}(z,w) f(w) d\mu(w) = 0,$$

which contradicts (1.4.12). This establishes (1.4.4). The identity (1.4.5) is a simple rearrangement of (1.4.4), and we end the proof of the lemma.

We are now at the final stage of the proof of Theorem 1.1: Let f be an arbitrary element of $B^\infty(\mathfrak{F})$. We put

$$f^* = \prod_{\nu=0}^{N}(\Delta + (\alpha+\nu)(\alpha+\nu-1))f,$$

where α and N are as in (1.4.3). Then we have, by (1.3.3) and (1.4.5),

$$f(z) = \sum_{j=0}^{\infty} q_\alpha^*(\lambda_j)\langle f^*, \psi_j\rangle \psi_j(z) + \frac{1}{4\pi}\int_{-\infty}^{\infty} q_\alpha^*(t^2+\tfrac{1}{4})\mathcal{E}(t,f^*)E(z,\tfrac{1}{2}+it)dt,$$

where $\mathcal{E}$ is as in (1.1.46), and the convergence is obviously absolute and uniform. But (1.1.28) gives

$$\langle f^*, \psi_j\rangle = \langle f, \psi_j^*\rangle = q_\alpha^*(\lambda_j)^{-1}\langle f, \psi_j\rangle;$$

and a multiple use of Green's formula gives

$$\mathcal{E}(t,f^*) = q_\alpha^*(t^2+\tfrac{1}{4})^{-1}\mathcal{E}(t,f). \tag{1.4.17}$$

Thus we have, in $B^\infty(\mathfrak{F})$,

$$f(z) = \sum_{j=0}^{\infty} \langle f, \psi_j\rangle \psi_j(z) + \frac{1}{4\pi}\int_{-\infty}^{\infty} \mathcal{E}(t,f)E(z,\tfrac{1}{2}+it)dt. \tag{1.4.18}$$

Since $\mathcal{E}(t,f) \in L^2(0,\infty)$ by (1.4.17), the assertions (1.1.44) and (1.1.45) have been almost proved.

What remains for us to do is to see the nature of the space $\mathfrak{E}$. For this purpose we show

Lemma 1.11 *Let $p \in L^2(0,\infty)$. Then*

$$\operatorname*{l.i.m.}_{T\to\infty} \int_0^T p(t)E(z,\tfrac{1}{2}+it)dt$$

exists in the space $L^2(\mathfrak{F},d\mu)$; that is, we have $\mathfrak{E} \subseteq L^2(\mathfrak{F},d\mu)$. Denoting this limit function by $P(z)$ we have

$$\langle P, \psi_j\rangle = 0 \quad (j=0,1,\dots) \tag{1.4.19}$$

and

$$\int_{\mathfrak{F}} |P(z)|^2 d\mu(z) = 2\pi\int_0^\infty |p(t)|^2 dt. \tag{1.4.20}$$

Proof Obviously it is sufficient to prove the last two identities when p is a C^∞-function having a compact support contained in the interval $(0, U)$, say, where $U > 0$ is arbitrary; needless to say, all estimates below may depend on U. Then the proof of (1.4.19) is reduced to (1.3.54) if $j \geq 1$; but the case $j = 0$ requires additional consideration, and is to be treated below.

We have, by (1.1.9),

$$\begin{aligned} P(z) &= \int_0^\infty p(t)E(z, \tfrac{1}{2} + it)dt \\ &= y^{\frac{1}{2}} \int_0^\infty p(t)y^{it}dt + y^{\frac{1}{2}} \int_0^\infty p(t)\varphi_\Gamma(\tfrac{1}{2} + it)y^{-it}dt + O(e^{-y}). \end{aligned} \tag{1.4.21}$$

The ordinary Parseval formula gives

$$\begin{aligned} \int_{\mathcal{F}} \left| y^{\frac{1}{2}} \int_0^\infty p(t)y^{it}dt \right|^2 d\mu(z) &\leq \int_0^\infty \left| \int_0^\infty p(t)e^{it \log y}dt \right|^2 y^{-1}dy \\ &= 2\pi \int_0^\infty |p(t)|^2 dt. \end{aligned}$$

Estimating the contribution of other parts in (1.4.21) similarly, we get $P \in L^2(\mathcal{F}, d\mu)$. Having this we can treat the case $j = 0$ in (1.4.19). We have

$$\begin{aligned} \langle P, 1 \rangle &= \lim_{Y \to \infty} \int_{\mathcal{F}_Y} P(z)d\mu(z) \\ &= \lim_{Y \to \infty} \int_0^\infty p(t) \int_{\mathcal{F}_Y} E(z, \tfrac{1}{2} + it)d\mu(z)dt \\ &= \lim_{Y \to \infty} \int_0^\infty \frac{p(t)}{t^2 + \frac{1}{4}} \int_{\mathcal{F}_Y} \Delta E(z, \tfrac{1}{2} + it)d\mu(z)dt. \end{aligned}$$

But the last integral is $O(Y^{-\frac{1}{2}})$ by Green's formula. This ends the proof of (1.4.19).

Turning to (1.4.20) we have

$$\begin{aligned} \|P\|^2 &= \lim_{Y \to \infty} \int_{\mathcal{F}_Y} \left| \int_0^\infty p(t)E(z, \tfrac{1}{2} + it)dt \right|^2 d\mu(z) \\ &= \lim_{Y \to \infty} \int_0^\infty \int_0^\infty p(t_1)\overline{p(t_2)}[E](t_1, t_2; Y)dt_1 dt_2, \end{aligned}$$

where

$$[E](t_1, t_2; Y) = \int_{\mathcal{F}_Y} E_1 E_2 d\mu(z);$$

$$E_1 = E(z, \tfrac{1}{2} + it_1), \quad E_2 = E(z, \tfrac{1}{2} - it_2).$$

If $t_1 \neq \pm t_2$, then Green's formula gives

$$[E](t_1,t_2;Y) = (t_1^2 - t_2^2)^{-1} \int_{\mathcal{F}_Y} [E_2 \Delta E_1 - E_1 \Delta E_2] d\mu(z)$$
$$= -(t_1^2 - t_2^2)^{-1} \int_{-\frac{1}{2}}^{\frac{1}{2}} [E_2 \frac{\partial}{\partial y} E_1 - E_1 \frac{\partial}{\partial y} E_2] dx \quad (y = Y).$$

The last integral is equal to

$$i(t_1+t_2)Y^{i(t_1-t_2)} + i(t_1-t_2)\varphi_\Gamma(\tfrac{1}{2} - it_2)Y^{i(t_1+t_2)}$$
$$- i(t_1-t_2)\varphi_\Gamma(\tfrac{1}{2} + it_1)Y^{-i(t_1+t_2)}$$
$$- i(t_1+t_2)\varphi_\Gamma(\tfrac{1}{2} + it_1)\varphi_\Gamma(\tfrac{1}{2} - it_2)Y^{-i(t_1-t_2)}$$
$$+ \int_{-\frac{1}{2}}^{\frac{1}{2}} [\tilde{E}_2 \frac{\partial}{\partial y} \tilde{E}_1 - \tilde{E}_1 \frac{\partial}{\partial y} \tilde{E}_2] dx,$$

where we have used (1.3.36) with an obvious abbreviation. Using Green's formula in the reverse way, we see that the last term is equal to

$$\int_{\mathcal{F}\setminus\mathcal{F}_Y} [\tilde{E}_2 \Delta \tilde{E}_1 - \tilde{E}_1 \Delta \tilde{E}_2] d\mu(z) = (t_1^2 - t_2^2) \int_{\mathcal{F}\setminus\mathcal{F}_Y} \tilde{E}_1 \tilde{E}_2 d\mu(z),$$

since we have (1.3.42). From these we get

$$[E](t_1,t_2;Y) = \frac{Y^{i(t_1-t_2)} - Y^{-i(t_1-t_2)}}{i(t_1-t_2)}$$
$$+ \frac{Y^{i(t_2-t_1)}}{i(t_1-t_2)} (1 - \varphi_\Gamma(\tfrac{1}{2} + it_1)\varphi_\Gamma(\tfrac{1}{2} - it_2))$$
$$- \frac{Y^{-i(t_1+t_2)}}{i(t_1+t_2)} \varphi_\Gamma(\tfrac{1}{2} + it_1) + \frac{Y^{i(t_1+t_2)}}{i(t_1+t_2)} \varphi_\Gamma(\tfrac{1}{2} - it_2)$$
$$- \int_{\mathcal{F}\setminus\mathcal{F}_Y} \tilde{E}_1 \tilde{E}_2 d\mu(z).$$

This holds for any real t_1, t_2 because of continuity. Thus, applying integration by parts appropriately, we have

$$\int_0^\infty \int_0^\infty p(t_1)\overline{p(t_2)}[E](t_1,t_2;Y) dt_1 dt_2$$
$$= 2 \int_0^\infty \int_0^\infty p(t_1)\overline{p(t_2)} \frac{\sin((t_1-t_2)\log Y)}{t_1 - t_2} dt_1 dt_2 + O((\log Y)^{-1})$$
$$= 2 \int_0^\infty \int_0^\infty |p(t_1)|^2 \frac{\sin((t_1-t_2)\log Y)}{t_1 - t_2} dt_1 dt_2 + O((\log Y)^{-1})$$
$$= 2\pi \int_0^\infty |p(t)|^2 dt + O((\log Y)^{-1}),$$

which gives rise to (1.4.20). This ends the proof of the lemma.

We are now ready to finish the proof of Theorem 1.1 : The last lemma and the assertion (1.4.18) yield

$$B^\infty(\mathfrak{F}) \subseteq \mathbb{C} \oplus \mathfrak{D} \oplus \mathfrak{E} \subseteq L^2(\mathfrak{F}, d\mu),$$

which proves the spectral decomposition (1.1.44) since $B^\infty(\mathfrak{F})$ is dense in $L^2(\mathfrak{F}, d\mu)$. We then pick out an arbitrary $f \in L^2(\mathfrak{F}, d\mu)$. By what we have just proved, we have that in the sense of the norm convergence

$$f(z) = \sum_{j=0}^{\infty} \omega_j \psi_j(z) + \int_0^\infty p(t) E(z, \tfrac{1}{2} + it) dt; \tag{1.4.22}$$

and hence

$$\|f\|^2 = \sum_{j=0}^{\infty} |\omega_j|^2 + 2\pi \int_0^\infty |p(t)|^2 dt \tag{1.4.23}$$

with certain complex numbers ω_j and a $p \in L^2(0, \infty)$. It is trivial to see that

$$\omega_j = \langle f, \psi_j \rangle \quad (j = 0, 1, \dots).$$

To fix p in (1.4.22) we introduce the Γ-automorphic truncation f_Y of f that is equal to $f(z)$ if $z \in \mathfrak{F}_Y$ and to 0 at other points in $\mathfrak{F}$. We consider the inner product of f_Y with the function in $\mathfrak{E}$ corresponding to the characteristic function of an arbitrary interval $[\tau_1, \tau_2]$ on the positive real axis: We have, in view of (1.4.20),

$$\int_{\mathfrak{F}_Y} f(z) \int_{\tau_1}^{\tau_2} E(z, \tfrac{1}{2} - it) dt d\mu(z) = 2\pi \int_{\tau_1}^{\tau_2} p_Y(t) dt,$$

where p_Y appears in place of p when the expansion (1.4.22) is applied to f_Y. The order of integration on the left side can of course be exchanged; and we find that for almost all t

$$p_Y(t) = \frac{1}{2\pi} \int_{\mathfrak{F}_Y} f(z) E(z, \tfrac{1}{2} - it) d\mu(z).$$

On the other hand, replacing f in (1.4.23) by $f - f_Y$ we get

$$2\pi \int_0^\infty |p(t) - p_Y(t)|^2 dt \le \|f - f_Y\|^2,$$

which amounts to

$$p(t) = \mathcal{E}(t, f).$$

Therefore we have established (1.1.45). This ends the proof of Theorem 1.1.

1.5 Notes for Chapter 1

The concept of real-analytic cusp-forms as eigenfunctions of Δ was introduced by Maass [41]; and thus those special automorphic functions are called *Maass waves.* His seminal work is an outcome of the rich tradition started by Hecke and Petersson. As we shall see in Chapter 3, Hecke [17] investigated the relation between holomorphic cusp-forms and Dirichlet series with Euler products, and his theory was greatly advanced by the metric theory created by Petersson [59] (see Section 2.2 below). Maass extended these to the real-analytic situation, being followed by Roelcke [63] and Selberg [65] who drew a fascinating picture of the spectral resolution of Δ in terms of Maass waves and Eisenstein series. They initiated extraordinary developments in the theory of automorphic functions, which now belong to the most important implements in modern number theory.

Since we are exclusively concerned with the application to the Riemann zeta-function, we restricted ourselves to the full modular group. It should be stressed, however, that our argument works for more general situations provided the corresponding Eisenstein series are similar to the above $E(z,s)$ in analytic characteristics. For instance, congruence subgroups can be treated completely analogously although there are additional complexities related to the existence of non-congruent cusps. Also, in these arithmetic cases we can add enhancements such as the introduction of multipliers and arbitrary real weights. In the latter extension the spectral theory is controlled by the differential operator of the form $\Delta + \nu i y\, \partial/\partial x$, and our argument still works well if appropriate modifications of simple nature are employed. Another direction of extension includes Hilbert's modular groups, which yields a possibility of new analytic investigations of algebraic number fields. At any event readers will not have difficulties in studying more general situations, provided they have the background developed in the above.

The literature of the theory of real-analytic automorphic functions is formidable. Moreover, save for Hejhal [18] and a few others, most books and articles are either too general or pedantic, lacking details and sometimes even proofs; thus defying beginners. Hejhal's volume is recommendable to the patient readers who may want to understand rigorously the generic situation. The geometric background is amply supplied in Siegel [67], while the analytical tools such as the Hilbert–Schmidt theory can be found in Courant and Hilbert [7] and in Riesz and Sz.-Nagy [62], though we need only the very basics. Elements of the Bessel and the hypergeometric functions are compactly developed in Lebedev [38]. The Mellin–Barnes formula for the hypergeometric function can be found in Titchmarsh [69] and in Whittaker and Watson [75].

As to the history of the non-Euclidean geometry, Milnor [43] is a readable account. The concept of non-Euclidean geometry was, as is well-known, first published by Lobachevsky. It appears that the model $\mathcal{H}$ with the metric given above and its extension to the higher dimensional situation had evolved in works due to Gauss, Riemann, and Liouville, before the explicit implementation by Beltrami.

Here are some minor remarks: The proof of the lower bound (1.1.34) is due to Roelcke [63]. For further numerical results see, e.g., Sarnak [64]. Our expression (1.1.49) for the free-space resolvent kernel differs from the one common in the literature; but they are actually the same, since

$$F(a,b;c;z) = (1-z)^{-a}F(a,c-b;c;z/(z-1)) \quad \left(|\arg(1-z)| < \pi\right)$$

(see [38, p. 247]). We employed (1.1.49) because it allows us to use the representation (1.2.2) whose integrand is of exponential decay, and thus easy to handle.

The titles of the second and the third sections are somewhat misleading. For we have actually avoided exploiting the fact that $\mathcal{R}_\alpha$ is the resolvent of $\Delta + \alpha(\alpha-1)$, i.e.,

$$(\Delta + \alpha(\alpha-1))\mathcal{R}_\alpha f = f$$

for sufficiently smooth f. Also we have not positively mentioned Hilbert's resolvent identity

$$\mathcal{R}_\alpha - \mathcal{R}_\beta = (\alpha(1-\alpha) - \beta(1-\beta))\mathcal{R}_\alpha\mathcal{R}_\beta$$

in treating the transformation $\mathcal{R}_{\alpha,N}$. It is behind our argument. Thus we have in fact

$$\mathcal{R}_{\alpha,N} = \mathcal{R}_\alpha\mathcal{R}_{\alpha+1}\cdots\mathcal{R}_{\alpha+N}.$$

The proof of the first among the last three equations is, however, not simple (see, e.g., Iwaniec [27, pp. 36–37]); and we think it is excessive in so far as we are concerned with the proof of the spectral resolution of Δ for arithmetic cases. In this context it should be stressed that the equation in question is immediate once we have established the spectral resolution of Δ; hence we have implicitly proved in the above that $\mathcal{R}_\alpha$ is indeed the resolvent.

In Section 1.4 we used some crude uniform bounds, which are by no means the best that we can prove but sufficient for our purpose. For example, the condition (1.4.3) could be much relaxed; and the bound (1.4.8) could be replaced by a sharper one. For the latter subject see Iwaniec [27, Chapter 13] as well as Sarnak [64].

2
Trace formulas

The aim of the present chapter is to prove transformation formulas of the sums

$$K_{\pm}(m,n\,;\varphi) = \sum_{l=1}^{\infty} \frac{1}{l} S(m, \pm n\,; l)\varphi\Big(\frac{4\pi}{l}\sqrt{mn}\Big),$$

where m, n are arbitrary positive integers, φ a smooth function satisfying certain decay conditions, and $S(m,n\,;l)$ the Kloosterman sum defined by (1.1.7). The results are expressed in terms of spectral sums of Fourier coefficients of holomorphic and real-analytic cusp-forms over the full modular group Γ. Thus they can be regarded as trace formulas. The theory was created by N.V. Kuznetsov; but what we are elaborating below is an alternative approach to his results with our own twists. Expressions of the $K_{\pm}$ type will appcar in our investigation on the Riemann zeta-function to be developed in later chapters. We shall retain the basic notations introduced in the previous chapter.

2.1 Basic identities

The key idea, which is due to A. Selberg, is to transform the inner-product of two Poincaré series,

$$\langle P_m(\cdot, s_1),\, P_n(\cdot, \overline{s_2})\rangle = \int_{\mathcal{F}} P_m(z, s_1)\overline{P_n(z, \overline{s_2})}d\mu(z),$$

in two ways: The definition of Poincaré series induces a geometrical argument – the unfolding method of Rankin and Selberg – which will lead us, via the formula (1.1.6), to an expression involving Kloosterman sums. On the other hand the Parseval formula (1.1.47) yields an expression involving Maass–Fourier coefficients. Thus, equating these expressions, we shall get a spectral expansion of a sum of Kloosterman sums, which is a precursor of the trace formulas.

The result of the application of the unfolding method is stated in

Lemma 2.1 *If*

$$\operatorname{Re} s_2 + \alpha > \operatorname{Re} s_1 > \alpha + \tfrac{3}{4} \quad (\alpha > 0), \tag{2.1.1}$$

then we have, for any integers $m, n \geq 1$,

$$\begin{aligned}\langle P_m(\cdot, s_1), P_n(\cdot, \overline{s_2})\rangle &= \delta_{m,n}\Gamma(s_1 + s_2 - 1)(4\pi m)^{1-s_1-s_2} \\ &+ 2^{2(1-s_2)}\pi^{s_1-s_2+1}n^{s_1-s_2}\Gamma(s_1 + s_2 - 1) \\ &\times \sum_{l=1}^{\infty} l^{-2s_1} S(m, n\,; l) W\left(\frac{4\pi}{l}\sqrt{mn}\,; s_1, s_2\right),\end{aligned} \tag{2.1.2}$$

where $\delta_{m,n}$ *is the Kronecker delta, and*

$$W(x\,; s_1, s_2) = \frac{1}{2\pi i}\int_{(\alpha)} \frac{\Gamma(\eta)\Gamma(\eta + s_2 - s_1)}{\Gamma(\eta + s_2)\Gamma(s_1 - \eta)}\left(\frac{x}{2}\right)^{-2\eta} d\eta. \tag{2.1.3}$$

Proof We need first to bound $P_m(z, s)$. For this purpose we quote the well-known estimate

$$|S(m, n\,; l)| \leq d(l)(m, n, l)^{\frac{1}{2}} l^{\frac{1}{2}}, \tag{2.1.4}$$

which implies that the double sum in (1.1.6) converges absolutely if

$$\operatorname{Re} s > \tfrac{3}{4}, \quad m \geq 1. \tag{2.1.5}$$

On this condition $P_m(z, s)$ is a regular function of s, and satisfies the bound

$$P_m(z, s) \ll y^{1-\operatorname{Re} s} \tag{2.1.6}$$

if y is not too small. Thus $P_m(z, s)$ belongs to $L^2(\mathcal{F}, d\mu)$ given (2.1.5).

Let f be in $L^2(\mathcal{F}, d\mu)$; and let $\operatorname{Re} s > 1$, $m \geq 1$. Then we have

$$\begin{aligned}\langle f, P_m(\cdot, \bar{s})\rangle &= \sum_{\gamma \in \Gamma_\infty \backslash \Gamma} \int_{\mathcal{F}} f(z)(\operatorname{Im} \gamma(z))^s e(-m\gamma(\bar{z}))d\mu(z) \\ &= \sum_{\gamma \in \Gamma_\infty \backslash \Gamma} \int_{\gamma(\mathcal{F})} f(z)(\operatorname{Im} z)^s e(-m\bar{z})d\mu(z),\end{aligned}$$

where the second line is due to the automorphy of f, and the first to the fact that by (1.1.9)

$$\sum_{\gamma\in\Gamma_\infty\backslash\Gamma} |\mathrm{Im}\,\gamma(z)|^\sigma |e(m\gamma(z))| \le E(z,\sigma) - y^\sigma + y^\sigma e^{-2\pi m y} \ll y^{1-\sigma}$$

$$(z \in \mathcal{F}, \sigma > 1).$$

Then we observe that if the representatives $\gamma \in \Gamma_\infty \setminus \Gamma$ are chosen appropriately the tiles $\gamma(\mathcal{F})$ fill up the domain $\{z \in \mathcal{H} : -\frac{1}{2} \le x \le \frac{1}{2}\}$ without any overlap of positive measure. Hence we get

$$\langle f, P_m(\cdot,\bar{s})\rangle = \int_0^\infty \int_{-\frac{1}{2}}^{\frac{1}{2}} f(z) e(-m\bar{z}) y^{s-2} dx dy. \tag{2.1.7}$$

This procedure is the unfolding method.

In particular, if $\mathrm{Re}\, s_1 > \frac{3}{4}$ and $\mathrm{Re}\, s_2 > 1$ then

$$\langle P_m(\cdot,s_1), P_n(\cdot,\overline{s_2})\rangle = \int_0^\infty \int_{-\frac{1}{2}}^{\frac{1}{2}} P_m(z,s_1) e(-n\bar{z}) y^{s_2-2} dx dy.$$

Thus the expansion (1.1.6) yields, for $\mathrm{Re}\, s_j > 1$ $(j = 1,2)$,

$$\begin{aligned}
&\langle P_m(\cdot,s_1), P_n(\cdot,\overline{s_2})\rangle \\
&= \delta_{m,n} \Gamma(s_1+s_2-1)(4\pi m)^{1-s_1-s_2} + \int_0^\infty y^{s_2-s_1-1} \sum_{l=1}^\infty l^{-2s_1} S(m,n\,;l) \\
&\times \int_{-\infty}^\infty \exp\Big(-2\pi n y(1+i\xi) - \frac{2\pi m}{l^2 y(1-i\xi)}\Big) \frac{d\xi}{(1+\xi^2)^{s_1}} dy,
\end{aligned} \tag{2.1.8}$$

which is to be transformed into (2.1.2). To this end we impose the condition

$$\sigma_2 > \sigma_1 > 1, \tag{2.1.9}$$

where $\mathrm{Re}\, s_j = \sigma_j$. This ensures absolute convergence on the right side of (2.1.8); and we may exchange freely the order of sums and integrals, getting

$$\begin{aligned}
&\langle P_m(\cdot,s_1),\ P_n(\cdot,\overline{s_2})\rangle \\
&\quad = \delta_{m,n} \Gamma(s_1+s_2-1)(4\pi m)^{1-s_1-s_2} \\
&\quad + \sum_{l=1}^\infty l^{-2s_1} S(m,n\,;l) \int_{-\infty}^\infty (1+\xi^2)^{-s_1} Y_{s_2-s_1}(\xi\,;m,n,l) d\xi,
\end{aligned} \tag{2.1.10}$$

where

$$Y_\omega(\xi\,;m,n,l) = \int_0^\infty y^{\omega-1} \exp\Big(-2\pi n y(1+i\xi) - \frac{2\pi m}{l^2 y(1-i\xi)}\Big) dy.$$

Then Mellin's formula gives

$$Y_\omega(\xi\,;m,n,l) = \frac{1}{2\pi i}\int_0^\infty y^{\omega-1}\exp(-2\pi ny(1+i\xi)) \times \int_{(\alpha)} \Gamma(\eta)\Big(\frac{2\pi m}{l^2 y(1-i\xi)}\Big)^{-\eta} d\eta dy,$$

where $\alpha > 0$ is arbitrary and $|\arg(1-i\xi)| < \frac{1}{2}\pi$. This double integral converges absolutely if

$$\operatorname{Re}\omega > -\alpha.$$

We exchange the order of integration and compute the resulting inner integral, getting

$$Y_\omega(\xi\,;m,n,l) = \frac{1}{2\pi i}(2\pi n)^{-\omega}\int_{(\alpha)}\Gamma(\eta)\Gamma(\eta+\omega)\Big(\frac{2\pi}{l}\sqrt{mn}\Big)^{-2\eta}\frac{(1-i\xi)^\eta}{(1+i\xi)^{\eta+\omega}}d\eta,$$

where $|\arg(1+i\xi)| < \frac{1}{2}\pi$. Thus, providing

$$\alpha + \sigma_2 - \sigma_1 > 0, \tag{2.1.11}$$

we have

$$\int_{-\infty}^\infty (1+\xi^2)^{-s_1} Y_{s_2-s_1}(\xi\,;m,n,l)d\xi = \frac{1}{2\pi i}(2\pi n)^{s_1-s_2} \times \int_{-\infty}^\infty\int_{(\alpha)}\Gamma(\eta)\Gamma(\eta+s_2-s_1)\Big(\frac{2\pi}{l}\sqrt{mn}\Big)^{-2\eta}\frac{(1-i\xi)^{\eta-s_1}}{(1+i\xi)^{\eta+s_2}}d\eta d\xi.$$

This double integral converges absolutely if

$$\alpha - \sigma_1 + \tfrac{1}{2} < 0, \tag{2.1.12}$$

since the integrand is, by Stirling's formula,

$$\begin{aligned}&\ll |\xi|^{-\sigma_1-\sigma_2}|\eta|^{2\alpha-\sigma_1+\sigma_2-1}\\ &\qquad\times\exp\big(-\pi|\eta| + (\operatorname{Im}\eta)\arg(1+i\xi) - (\operatorname{Im}\eta)\arg(1-i\xi)\big)\\ &\ll |\xi|^{-\sigma_1-\sigma_2}|\eta|^{2\alpha-\sigma_1+\sigma_2-1}\exp\big(-|\eta|/|\xi|\big),\end{aligned}$$

provided $|\eta|$ and $|\xi|$ are sufficiently large. Thus, given (2.1.12), we may exchange the order of integration. In the resulting inner integral we deform the contour and see that it is equal to

$$\int_C \frac{(1-i\xi)^{\eta-s_1}}{(1+i\xi)^{\eta+s_2}}d\xi,$$

where C starts at $+i\infty$, goes down along the imaginary axis, describes once a small circle round the point i in the positive direction, and returns to $+i\infty$. We note that $1-i\xi$ remains essentially positive and $\arg(1+i\xi)$ varies from $-\pi$ to π round the contour. We assume temporarily that $\alpha+\sigma_2<1$. Then the circular part of C can be collapsed to the point i, and the integral is equal to

$$\begin{aligned}&\left(-ie^{\pi i(\eta+s_2)}+ie^{-\pi i(\eta+s_2)}\right)\int_1^\infty \frac{(1+\xi)^{\eta-s_1}}{(\xi-1)^{\eta+s_2}}d\xi\\ &=2\sin(\pi(\eta+s_2))2^{1-s_1-s_2}\Gamma(1-\eta-s_2)\Gamma(s_1+s_2-1)/\Gamma(s_1-\eta)\\ &=2^{2-s_1-s_2}\pi\Gamma(s_1+s_2-1)/\{\Gamma(\eta+s_2)\Gamma(s_1-\eta)\}.\end{aligned}$$

By analytic continuation we may obviously drop the interim condition. Combining these identities we end the proof of the lemma.

It should be noted that (2.1.1) means that the conditions (2.1.11) and (2.1.12) are fulfilled, and (2.1.9) has been removed. The identity (2.1.2) could be rearranged so that the symmetry with respect to the parameters s_1, s_2, m, n became visible on the right side too; i.e., replace η in (2.1.3) by $\eta+\frac{1}{2}(s_1-s_2)$. We also make the observation that the integrand in (2.1.3) reduces to simple expressions either when $s_1+s_2=2$ or when $s_1=1$. This fact will be exploited in later sections.

We next state the spectral expansion of the inner-product of two Poincaré series:

Lemma 2.2 *Let* $\{\psi_j : j\geq 1\}$ *be as in* (1.1.41) *and* (1.1.43). *Then, providing*

$$m,\, n\geq 1,\quad \operatorname{Re} s_j>\tfrac{3}{4}\quad (j=1,2),\tag{2.1.13}$$

we have

$$\begin{aligned}&\langle P_m(\cdot,s_1),\, P_n(\cdot,\overline{s_2})\rangle\\ &=\frac{\pi}{\Gamma(s_1)\Gamma(s_2)}(4\pi\sqrt{mn})^{1-s_1-s_2}(n/m)^{\frac{1}{2}(s_1-s_2)}\\ &\times\Bigg\{\sum_{j=1}^\infty \overline{\rho_j(m)}\rho_j(n)\Theta(s_1,s_2\,;\kappa_j)\\ &+\frac{1}{\pi}\int_{-\infty}^\infty \frac{\sigma_{2ir}(m)\sigma_{2ir}(n)\cosh(\pi r)\Theta(s_1,s_2\,;r)}{(mn)^{ir}|\zeta(1+2ir)|^2}dr\Bigg\},\end{aligned}\tag{2.1.14}$$

where

$$\Theta(s_1, s_2; r) = \Gamma(s_1 - \tfrac{1}{2} + ir)\Gamma(s_1 - \tfrac{1}{2} - ir)\Gamma(s_2 - \tfrac{1}{2} + ir)\Gamma(s_2 - \tfrac{1}{2} - ir). \quad (2.1.15)$$

Proof By the observation made at the beginning of the proof of the previous lemma both $P_m(z, s_1)$ and $P_n(z, s_2)$ are in $L^2(\mathcal{F}, d\mu)$ on the assumption (2.1.13). Thus the Parseval formula (1.1.47) yields that, given (2.1.13),

$$\langle P_m(\cdot, s_1), P_n(\cdot, \overline{s_2})\rangle = \sum_{j=0}^{\infty} \langle P_m(\cdot, s_1), \psi_j\rangle \overline{\langle P_n(\cdot, \overline{s_2}), \psi_j\rangle} + \frac{1}{4\pi}\int_{-\infty}^{\infty} \mathcal{E}(r, P_m(\cdot, s_1))\overline{\mathcal{E}(r, P_n(\cdot, \overline{s_2}))}\, dr. \quad (2.1.16)$$

The formula (2.1.7) gives, for $\operatorname{Re} s > 1$,

$$\langle P_m(\cdot, s), \psi_j\rangle = \int_0^{\infty}\int_{-\frac{1}{2}}^{\frac{1}{2}} \overline{\psi_j(z)} y^{s-2} e(mz)\, dx\, dy \quad (j \ge 0).$$

This readily implies

$$\langle P_m(\cdot, s), \psi_0\rangle = 0. \quad (2.1.17)$$

Also, if $j \ge 1$ then we have, by (1.1.41),

$$\begin{aligned}\langle P_m(\cdot, s), \psi_j\rangle &= (2\pi m)^{\frac{1}{2}-s}\overline{\rho_j(m)} \int_0^{\infty} y^{s-\frac{3}{2}} e^{-y} K_{i\kappa_j}(y)\, dy \\ &= \sqrt{\pi}(4\pi m)^{\frac{1}{2}-s}\overline{\rho_j(m)}\Gamma(s - \tfrac{1}{2} + i\kappa_j)\Gamma(s - \tfrac{1}{2} - i\kappa_j)/\Gamma(s).\end{aligned} \quad (2.1.18)$$

In much the same way we see that (1.1.9) gives, for $\operatorname{Re} s > 1$, $r \in \mathbb{R}$,

$$\mathcal{E}(r, P_m(\cdot, s)) = 2^{2-2s}\pi(m\pi)^{\frac{1}{2}-s-ir}\sigma_{2ir}(m)\frac{\Gamma(s - \frac{1}{2} + ir)\Gamma(s - \frac{1}{2} - ir)}{\Gamma(s)\Gamma(\frac{1}{2} - ir)\zeta(1 - 2ir)}. \quad (2.1.19)$$

But if $\operatorname{Re} s > \frac{3}{4}$ then both sides of each of (2.1.17)–(2.1.19) are regular functions of s; thus they hold for $\operatorname{Re} s > \frac{3}{4}$ by analytic continuation. We insert these into (2.1.16), and end the proof of the lemma.

It should be noted that the evaluation of the integral in (2.1.18) has been made by invoking the representation (1.1.17): In fact, providing

$$\operatorname{Re} s > |\operatorname{Re} \eta| + \tfrac{1}{2}, \quad (2.1.20)$$

we have

$$\int_0^\infty y^{s-\frac{3}{2}} e^{-y} K_\eta(y) dy$$
$$= \frac{1}{2}\int_0^\infty \xi^{\eta-1} \int_0^\infty y^{s-\frac{3}{2}} \exp\left(-\tfrac{1}{2}y(\xi^{\frac{1}{2}}+\xi^{-\frac{1}{2}})^2\right) dy d\xi$$
$$= 2^{s-\frac{3}{2}}\Gamma(s-\tfrac{1}{2})\int_0^\infty \xi^{s+\eta-\frac{3}{2}}(\xi+1)^{1-2s} d\xi$$
$$= (2\pi)^{\frac{1}{2}} 2^{-s}\Gamma(s+\eta-\tfrac{1}{2})\Gamma(s-\eta-\tfrac{1}{2})/\Gamma(s), \qquad (2.1.21)$$

where the absolute convergence, in the region (2.1.20), of the double integral is easy to check, and the last line depends on the duplication formula for the Γ-function.

2.2 Holomorphic forms

Next, we introduce some basic notions related to the spectral theory of holomorphic automorphic forms due to H. Petersson. Although the contribution from holomorphic cusp-forms to our theory of the zeta-function will eventually turn out to be marginal (see Lemma 5.1), we should treat the subject with respect because of its importance in itself as well as the fact that it is only after the full development of our argument that neglecting it becomes admissible.

So let f be a function, regular throughout $\mathcal{H}$, such that with a certain fixed positive integer k we have $f(\gamma(z)) = j(\gamma,z)^{2k} f(z)$ for all $\gamma \in \Gamma$, where $j(\gamma,z)$ is the denominator of the fractional linear transform $\gamma(z)$, so that $(d/dz)\gamma(z) = j(\gamma,z)^{-2}$. Then we call f a Γ-automorphic form of weight $2k$. Since such an f is of period 1, it can be expanded into a power series in $e(z)$; and if only terms with positive exponent appear, i.e., $f(i\infty)=0$, then we call f a holomorphic cusp-form of weight $2k$ over Γ. For each k the collection of such functions yields a linear set, which we denote by $\mathscr{C}_k(\Gamma)$. We then have

Theorem 2.1 *The set $\mathscr{C}_k(\Gamma)$ is a finite dimensional unitary space equipped with the inner-product*

$$\langle f_1, f_2\rangle_k = \int_{\mathcal{F}} f_1(z)\overline{f_2(z)} y^{2k} d\mu(z). \qquad (2.2.1)$$

Proof The convergence of the integral is trivial because the cusp-forms decay exponentially near the cusp. Thus it is sufficient to prove the finiteness of the dimension: Let f_1 be a non-zero element of $\mathscr{C}_k(\Gamma)$. We assume that f_1 has N zeros including multiplicity on the compactified Riemann surface $\overline{\mathfrak{F}} = \mathfrak{F} \cup \{\text{cusp}\}$. We pick up arbitrary elements f_ν $(2 \le \nu \le N+1)$ of $\mathscr{C}_k(\Gamma)$. Then, by a fundamental assertion in linear algebra, there exists a non-trivial constant vector $\{c_\nu\}$ such that

$$\sum_{\nu=1}^{N+1} c_\nu f_\nu / f_1$$

is regular on $\overline{\mathfrak{F}}$, i.e., regular throughout $\mathcal{H}$ and finite at the cusp. So this sum is harmonic and in $L^2(\mathfrak{F}, d\mu)$; thus it should be a constant, as we have seen in the proof of Lemma 1.4. This ends the proof of the theorem.

We make

$$\{\psi_{j,k} \ : 1 \le j \le \vartheta(k)\} \text{ an orthonormal basis of } \mathscr{C}_k(\Gamma). \tag{2.2.2}$$

The Fourier coefficients $\rho_{j,k}(n)$ of $\psi_{j,k}(z)$ are defined by the expansion

$$\psi_{j,k}(z) = \sum_{n=1}^{\infty} \rho_{j,k}(n) n^{k-\frac{1}{2}} e(nz). \tag{2.2.3}$$

They will play a similar rôle to the Maass–Fourier coefficients. Here the dimension $\vartheta(k)$ can be computed explicitly: We have

$$\vartheta(k) = \begin{cases} 0 & \text{if } k = 1, \\ [k/6] - 1 & \text{if } k \equiv 1 \bmod 6,\ k \ne 1, \\ [k/6] & \text{if } k \not\equiv 1 \bmod 6. \end{cases} \tag{2.2.4}$$

But we shall not need this result; the finiteness of $\vartheta(k)$ is sufficient for our purpose.

We are going to show an analogue of the combination of the last two lemmas for holomorphic cusp-forms: For this purpose we put, with arbitrary integers $m \ge 1$, $k > 1$,

$$U_m(z,k) = \sum_{\gamma \in \Gamma_\infty \backslash \Gamma} j(\gamma, z)^{-2k} e(m\gamma(z)).$$

The argument leading to (1.1.6) readily gives the Fourier expansion

$$\begin{aligned} U_m(z,k) =& e(mz) + \sum_{n=-\infty}^{\infty} e(nz) \sum_{l=1}^{\infty} l^{-2k} S(m,n\,;l) \\ &\times \int_{-\infty}^{\infty} \exp\Big(-2\pi i n(\xi + yi) - \frac{2\pi i m}{l^2(\xi + yi)}\Big)(\xi + yi)^{-2k} d\xi. \end{aligned}$$

If $n \le 0$ then this integral vanishes, as can be seen by shifting the path to $\mathrm{Im}\,\xi = +\infty$. If $n \ge 1$ then we shift the path to $\mathrm{Im}\,\xi = -\infty$. The computation of the residue at $\xi = -yi$ is reduced to the definition of the J-Bessel function; and we find that

$$U_m(z,k) = e(mz) + 2(-1)^k \pi m^{\frac{1}{2}-k} \sum_{n=1}^{\infty} n^{k-\frac{1}{2}} q_{m,n}(k) e(nz), \tag{2.2.5}$$

where

$$q_{m,n}(k) = \sum_{l=1}^{\infty} \frac{1}{l} S(m,n\,;l) J_{2k-1}\left(\frac{4\pi}{l}\sqrt{mn}\right). \tag{2.2.6}$$

These infinite sums converge absolutely for $k \ge 1$, since we have, uniformly in $k \ge 1$ and $x \ge 0$,

$$J_{2k-1}(x) \ll \min\{1, (\tfrac{1}{2}x)^{2k-1}/\Gamma(2k)\}. \tag{2.2.7}$$

Thus we may redefine $U_m(z,k)$ by (2.2.5) for all integral $k \ge 1$; then they are holomorphic cusp-forms of weight $2k$ (the case $k = 1$ is treated below).

Here the bound (2.2.7) is a simple consequence of the definition

$$J_{2k-1}(x) = \pi^{-1} \int_0^{\pi} \cos((2k-1)\theta - x\sin\theta) d\theta$$

as well as the representation

$$J_\nu(x) = \frac{1}{\sqrt{\pi}\Gamma(\nu+\frac{1}{2})} \left(\frac{x}{2}\right)^{\nu} \int_{-1}^{1} (1-y^2)^{\nu-\frac{1}{2}} \cos(xy)\, dy\,, \tag{2.2.8}$$

where $x > 0, \mathrm{Re}\,\nu > -\frac{1}{2}$, which can be proved by replacing $\cos(xy)$ by its Taylor series.

Just as in the real-analytic case, the inner-product $\langle U_m(\cdot\,;k), U_n(\cdot\,;k)\rangle_k$ can be transformed in two ways, either by the unfolding method or by the Parseval formula with respect to the system (2.2.2). It leads us to Petersson's trace formula:

Lemma 2.3 *Let* $\{\rho_{j,k}(n)\}$ *and* $q_{m,n}(k)$ *be defined by* (2.2.2)–(2.2.3) *and* (2.2.6), *respectively. Then we have, for any integers* $k, m, n \ge 1$,

$$q_{m,n}(k) = \frac{1}{2\pi}(-1)^{k-1}\delta_{m,n} + (-1)^k \frac{\pi}{(2k-1)} a(k) \sum_{j=1}^{\vartheta(k)} \overline{\rho_{j,k}(m)} \rho_{j,k}(n), \tag{2.2.9}$$

where $a(k) = 2^{1-4k}\pi^{-2k-1}\Gamma(2k)$. *In particular we have, uniformly in* k, m,

$$a(k)\sum_{j=1}^{\vartheta(k)} |\rho_{j,k}(m)|^2 \ll kd_3(m)m^{\frac{1}{2}}\log(2m). \tag{2.2.10}$$

Proof We consider (2.2.10) first. We have, by (2.1.4), (2.2.7), and (2.2.9),

$$\begin{aligned} a(k)k^{-1}\sum_{j=1}^{\vartheta(k)} |\rho_{j,k}(m)|^2 \ll 1 &+ \sum_{l\le 2\pi m} l^{-\frac{1}{2}}(m,l)^{\frac{1}{2}}d(l) \\ &+ \Gamma(2k)^{-1}(2\pi m)^{2k-1}\sum_{l\ge 2\pi m} l^{\frac{1}{2}-2k}(m,l)^{\frac{1}{2}}d(l). \end{aligned} \tag{2.2.11}$$

To estimate these sums over l we note that

$$n^\alpha = \sum_{r|n}\phi_\alpha(r); \quad \phi_\alpha(r) = r^\alpha\prod_{p|r}(1-p^{-\alpha}). \tag{2.2.12}$$

Then the second sum over l in (2.2.11) is

$$\begin{aligned} &\ll \sum_{r|m} r^{\frac{1}{2}-2k}\phi_{\frac{1}{2}}(r)d(r)\sum_{l\ge 2\pi m/r} l^{\frac{1}{2}-2k}d(l) \\ &\ll (2\pi m)^{\frac{3}{2}-2k}\log(2m)\sum_{r|m} r^{-1}\varphi_{\frac{1}{2}}(r)d(r) \\ &\ll d_3(m)(2\pi m)^{\frac{3}{2}-2k}\log(2m) \end{aligned}$$

uniformly in $k, m \ge 1$. In just the same way we can show that the other sum over l in (2.2.11) is $O(d_3(m)m^{\frac{1}{2}}\log(2m))$; and we get (2.2.10).

We next consider (2.2.9). Since the assertion is immediate when $k > 1$, it is sufficient to prove it only for $k = 1$. For this purpose we put, for a positive integer m,

$$U_m(z, 1; s) = \sum_{\gamma\in\Gamma_\infty\backslash\Gamma} j(\gamma, z)^{-2}|j(\gamma, z)|^{-2s}e(m\gamma(z)), \tag{2.2.13}$$

which converges absolutely for $\mathrm{Re}\, s > 0$. The argument leading to (1.1.6) yields

$$\begin{aligned} U_m(z, 1; s) = e(mz) &+ \sum_{l=1}^{\infty} l^{-2-2s}\sum_{n=-\infty}^{\infty} S(m, n; l)e(nz) \\ &\times \int_{-\infty}^{\infty}(\xi+iy)^{-2}(\xi^2+y^2)^{-s}\exp\Big(-2\pi in(\xi+yi) - \frac{2\pi im}{l^2(\xi+yi)}\Big)d\xi. \end{aligned} \tag{2.2.14}$$

Shifting the path to $\operatorname{Im}\xi = -\frac{1}{2}\operatorname{sgn}(n)y$ we see that the integral is $O(e^{3n\pi y})$ if $n < 0$ and $O(e^{n\pi y})$ if $n \geq 0$, uniformly for $\operatorname{Re} s > -\frac{1}{2}$. Hence (2.1.4) implies that $U_m(z, 1; s)$ is regular for $\operatorname{Re} s > -\frac{1}{4}$, and $U_m(i\infty, 1; s) = 0$. In particular $U_m(z, 1; 0) = U_m(z, 1)$ by the definition (2.2.5); and thus $U_m(z, 1)$ is a cusp-form of weight 2 as can be seen from the relation

$$U_m(\gamma(z), 1; s) = j(\gamma, z)^2 |j(\gamma, z)|^{2s} U_m(z, 1; s) \tag{2.2.15}$$

for any $\gamma \in \Gamma$. Here we could use the fact $\vartheta(1) = 0$ (see (2.2.4)). This would imply $U_m(z, 1) \equiv 0$ for all $m \geq 1$; and $q_{m,n}(1) = \delta_{m,n}/(2\pi)$ via (2.2.5), which is equivalent to (2.2.9), $k = 1$. But we have to avoid this argument for an obvious reason. Thus we proceed as follows:

We have, at all events,

$$\langle U_m(z, 1), U_n(z, 1)\rangle_1 = \sum_{j=1}^{\vartheta(1)} \langle U_m(z, 1), \psi_{j,1}(z)\rangle_1 \overline{\langle U_m(z, 1), \psi_{j,1}(z)\rangle_1} \tag{2.2.16}$$

for any integers $m, n \geq 1$. To compute the left side we introduce

$$\mathscr{U}_{m,n}(s_1, s_2) = \int_{\mathscr{F}} U_m(z, 1; s_1)\overline{U_n(z, 1; \overline{s_2})} y^{2+s_1+s_2} d\mu(z).$$

This is regular for $\operatorname{Re} s_1, \operatorname{Re} s_2 > -\frac{1}{4}$. If $\operatorname{Re} s_2 > 0$ then the unfolding method combined with (2.2.15) readily gives

$$\mathscr{U}_{m,n}(s_1, s_2) = \int_0^\infty \int_{-\frac{1}{2}}^{\frac{1}{2}} U_m(z, 1; s_1) e(-n\bar{z}) y^{s_1+s_2} dx dy;$$

and we have, by (2.2.14),

$$\begin{aligned}\mathscr{U}_{m,n}(s_1, s_2) &= \delta_{m,n}\Gamma(s_1 + s_2 + 1)(4\pi m)^{-s_1-s_2-1} \\ &\quad - \int_0^\infty y^{s_2-s_1-1} \sum_{l=1}^\infty l^{-2s_1-2} S(m, n; l) \\ &\quad \times \int_{-\infty}^\infty \exp\Big(-2\pi n y(1 + i\xi) - \frac{2\pi m}{l^2 y(1 - i\xi)}\Big) \frac{d\xi}{(1 + \xi^2)^{s_1}(1 - i\xi)^2} dy,\end{aligned}$$

which holds for $\operatorname{Re} s_2 > \operatorname{Re} s_1 > -\frac{1}{4}$ because of the absolute convergence. This is analogous to (2.1.8); and we can follow the argument leading to the assertion of Lemma 2.1. We then get

$$\begin{aligned}\mathscr{U}_{m,n}(s_1, s_2) &= \delta_{m,n}\Gamma(s_1 + s_2 + 1)(4\pi m)^{-1-s_1-s_2} \\ &\quad - 2^{-2s_2}\pi^{s_1-s_2+1} n^{s_1-s_2}\Gamma(s_1 + s_2 + 1) \\ &\quad \times \sum_{l=1}^\infty l^{-2s_1-2} S(m, n; l) W^*\Big(\frac{4\pi}{l}\sqrt{mn}; s_1, s_2\Big),\end{aligned}$$

where

$$W^*(x\,;s_1,s_2)=\frac{1}{2\pi i}\int_{(\alpha)}\frac{\Gamma(\eta)\Gamma(\eta+s_2-s_1)}{\Gamma(\eta+s_2)\Gamma(2+s_1-\eta)}\left(\frac{x}{2}\right)^{-2\eta}d\eta \quad (\alpha>0).$$

This is valid for $\mathrm{Re}\, s_2 > \mathrm{Re}\, s_1 - \alpha > -\frac{1}{4}$. We note that we have

$$W^*(x;0,0)=2x^{-1}J_1(x).$$

For we have the representation

$$J_{2\nu}(x)=\frac{1}{2\pi i}\int_{(\beta)}\frac{\Gamma(\eta-\frac{1}{2}+\nu)}{\Gamma(\frac{3}{2}-\eta+\nu)}\left(\frac{x}{2}\right)^{1-2\eta}d\eta\,, \tag{2.2.17}$$

where $x>0$ and $\frac{1}{2}-\mathrm{Re}\,\nu<\beta<\frac{1}{2}$. This can be proved by shifting the contour to $-\infty$. Hence, putting $s_1=s_2=0$ in the above, we find that

$$\langle U_m(z,1),U_n(z,1)\rangle_1=\delta_{m,n}(4\pi m)^{-1}-(2\sqrt{mn})^{-1}q_{m,n}(1). \tag{2.2.18}$$

To compute the factors on the right side of (2.2.16) we consider

$$\int_{\mathcal{F}} U_m(z,1;s)\overline{\psi_{j,1}(z)}y^{2+s}d\mu(z),$$

which is regular for $\mathrm{Re}\, s>-\frac{1}{4}$. We see, by the unfolding method, that this is equal to

$$\int_0^\infty\int_{-\frac{1}{2}}^{\frac{1}{2}}\overline{\psi_{j,1}(z)}e(mz)y^s dxdy=(4\pi)^{-s-1}m^{-s-\frac{1}{2}}\Gamma(s+1)\overline{\rho_{j,1}(m)}$$

provided $\mathrm{Re}\, s>0$. Thus, by analytic continuation, we get

$$\langle U_m(z,1),\psi_{j,1}(z)\rangle_1=(4\pi\sqrt{m})^{-1}\overline{\rho_{j,1}(m)}.$$

Inserting this and (2.2.18) into (2.2.16) we obtain (2.2.9) when $k=1$, which finishes the proof of the lemma.

2.3 Spectral means of Maass–Fourier coefficients

The aim of this section is to establish a trace formula, which relates the spectral sum

$$\sum_{j=1}^{\infty}\frac{\overline{\rho_j(m)}\rho_j(n)}{\cosh(\pi\kappa_j)}f(\kappa_j) \tag{2.3.1}$$

to Kloosterman sums, where f is to satisfy certain mild conditions. For this purpose we need to prove first a statistical estimate of the Maass–Fourier coefficients, which is a counterpart of the assertion (2.2.10) for real-analytic cusp-forms, and will play a crucial rôle at various places in our later discussion:

Lemma 2.4 *Let* $\{\rho_j(n)\}$ *be the Maass–Fourier coefficients defined by* (1.1.41) *and* (1.1.43). *Then we have, uniformly in* K, $m \geq 1$,

$$\sum_{K/2<\kappa_j\leq K} |\rho_j(m)|^2 e^{-\pi\kappa_j} \ll K^2 + d_3(m)m^{\frac{1}{2}}\log(2m). \tag{2.3.2}$$

Proof We put in (2.1.2) and (2.1.14)

$$s_1 = 1 + it,\ s_2 = 1 - it \quad (\,|\mathrm{Im}\, t| < \tfrac{1}{4}\,). \tag{2.3.3}$$

Then we have

$$\sum_{j=1}^{\infty} \frac{\overline{\rho_j(m)}\rho_j(n)}{\cosh(\pi\kappa_j)} p(t,\kappa_j) + \frac{1}{\pi}\int_{-\infty}^{\infty} \frac{\sigma_{2ir}(m)\sigma_{2ir}(n)}{(mn)^{ir}|\zeta(1+2ir)|^2} p(t,r)dr$$

$$= \delta_{m,n}\frac{t}{\pi^2\sinh(\pi t)} + \frac{2t}{\pi\sinh(\pi t)}\sum_{l=1}^{\infty}\frac{1}{l}S(m,n\,;l)\varpi\left(t, \frac{4\pi}{l}\sqrt{mn}\right), \tag{2.3.4}$$

where

$$p(t,r) = \frac{\cosh(\pi r)}{\cosh(\pi(r+t))\cosh(\pi(r-t))},$$

and

$$\varpi(t,x) = \frac{1}{2\pi i}\int_{(\alpha)} \frac{\sin(\pi\eta)}{\pi\eta}\Gamma(\eta+it)\Gamma(\eta-it)\left(\frac{x}{2}\right)^{1-2\eta} d\eta \tag{2.3.5}$$

with α satisfying

$$|\mathrm{Im}\, t| < \alpha < \tfrac{1}{4}. \tag{2.3.6}$$

We let t be real and $m = n$ in (2.3.4). We multiply both sides by the non-negative factor

$$\left\{\exp(-(t/K)^2) - \exp(-(2t/K)^2)\right\}(\sinh \pi t)/t$$

with a large parameter $K > 0$, and integrate with respect to t over the real axis. We get

$$\sum_{j=1}^{\infty} \frac{|\rho_j(m)|^2}{\cosh(\pi\kappa_j)} P^{(1)}(\kappa_j, K) + \int_{-\infty}^{\infty} \frac{|\sigma_{2ir}(m)|^2}{|\zeta(1+2ir)|^2} P^{(1)}(r,K)dr$$

$$= (2\pi^{\frac{3}{2}})^{-1}K + 2\pi^{-1}\sum_{l=1}^{\infty}\frac{1}{l}S(m,m\,;l)P^{(2)}(4\pi m/l, K), \tag{2.3.7}$$

where

$$P^{(1)}(r,K)=\int_{-\infty}^{\infty}p(t,r)\big(e^{-(t/K)^2}-e^{-(2t/K)^2}\big)(\sinh \pi t)\frac{dt}{t}, \qquad (2.3.8)$$

$$P^{(2)}(x,K)=\int_{-\infty}^{\infty}\varpi(t,x)\big(e^{-(t/K)^2}-e^{-(2t/K)^2}\big)dt. \qquad (2.3.9)$$

The exchange of the order of sums and integrals that is implicit on both sides of (2.3.7) is easy to verify.

On noting that all terms on the left side of (2.3.7) are non-negative, we have

$$\sum_{\frac{1}{2}K\le\kappa_j\le K}\frac{|\rho_j(m)|^2}{\cosh(\pi\kappa_j)}P^{(1)}(\kappa_j,K)$$
$$\ll K+\sum_{l=1}^{\infty}\frac{1}{l}|S(m,m\,;l)||P^{(2)}(4\pi m/l,K)|.$$

To get a lower bound of $P^{(1)}(r,K)$, we restrict the range of integration in (2.3.8) to the interval $[K/2,\,K]$. We have, uniformly for $\frac{1}{2}K\le r\le K$,

$$P^{(1)}(r,K)\gg K^{-1}\int_{K/2}^{K}\big(\cosh(\pi(t-r))\big)^{-1}dt\gg K^{-1}.$$

Thus we have

$$\sum_{\frac{1}{2}K\le\kappa_j\le K}\frac{|\rho_j(m)|^2}{\cosh(\pi\kappa_j)}\ll K^2+K\sum_{l=1}^{\infty}\frac{1}{l}|S(m,m\,;l)||P^{(2)}(4\pi m/l,K)|. \qquad (2.3.10)$$

To estimate $P^{(2)}(x,K)$ we have to transform it a little. For this purpose we note that the factor $\Gamma(\eta+it)\Gamma(\eta-it)$ in (2.3.5) is equal to

$$\Gamma(2\eta)\int_0^1 y^{\eta+it-1}(1-y)^{\eta-it-1}dy. \qquad (2.3.11)$$

Inserting it into (2.3.9), we get a triple integral which is absolutely convergent. Thus, performing the t-integral first, we have

$$P^{(2)}(x,K)=\frac{\sqrt{\pi}}{2\pi i}\int_{(\alpha)}\frac{\sin(\pi\eta)}{\pi\eta}\Gamma(2\eta)\Big(\frac{x}{2}\Big)^{1-2\eta}$$
$$\times\int_0^1(y(1-y))^{\eta-1}Q(y,K)dyd\eta,$$

where

$$Q(y,K)=K\exp\Big(-\Big(\frac{K}{2}\log\frac{y}{1-y}\Big)^2\Big)-\frac{1}{2}K\exp\Big(-\Big(\frac{K}{4}\log\frac{y}{1-y}\Big)^2\Big).$$

Hence we find that

$$P^{(2)}(x,K) = -\frac{x}{2\sqrt{\pi}}\int_0^1 (y(1-y))^{-1}Q(y,K)\mathrm{si}\Big(\frac{x}{2\sqrt{y(1-y)}}\Big)dy, \quad (2.3.12)$$

where

$$\mathrm{si}(u) = -\int_u^\infty \frac{\sin\xi}{\xi}d\xi.$$

In fact we have, for any positive u and $0 < a < \frac{1}{4}$,

$$\frac{1}{2\pi i}\int_{(a)} \frac{\sin(\pi\eta)}{\pi\eta}\Gamma(2\eta)u^{-2\eta}d\eta = -\frac{1}{\pi}\mathrm{si}(u),$$

as can be seen by shifting the contour to $-\infty$. So we have

$$P^{(2)}(x,K) = \frac{x}{2\sqrt{\pi}}\int_0^1 (y(1-y))^{-1}Q(y,K)\Big(\mathrm{si}(x) - \mathrm{si}\Big(\frac{x}{2\sqrt{y(1-y)}}\Big)\Big)dy.$$

Here we have used the fact that

$$\int_0^1 (y(1-y))^{-1}Q(y,K)dy = 0.$$

This is due to the identity

$$\int_0^1 (v(1-v))^{-1}\exp\Big(-\Big(\frac{K}{2}\log\frac{v}{1-v}\Big)^2\Big)dv$$
$$= \frac{1}{2}\int_0^1 (y(1-y))^{-1}\exp\Big(-\Big(\frac{K}{4}\log\frac{y}{1-y}\Big)^2\Big)dy,$$

which is a result of the change of variable $y/(1-y) = (v/(1-v))^2$ on the right side. Then, noting that $\mathrm{si}(u) \ll (1+u)^{-1}$ for non-negative u, we have

$$P^{(2)}(x,K) \ll x\int_0^1 (y(1-y))^{-1}|Q(y,K)|\min\Big((1+x)^{-1}, |\log(4y(1-y))|\Big)dy$$
$$\ll x\min\Big(\frac{1}{1+x}, \frac{1}{K^2}\Big) \ll \frac{x}{x+K^2}. \quad (2.3.13)$$

We insert this into (2.3.10) and invoke (2.1.4), getting

$$\sum_{K/2<\kappa_j\le K} |\rho_j(m)|^2 e^{-\pi\kappa_j} \ll K^2 + mK^{-1}\sum_{l=1}^\infty \frac{(l,m)^{\frac{1}{2}}d(l)}{l^{\frac{1}{2}}(l+mK^{-2})}.$$

Using the device (2.2.12) again, we see that the last sum is

$$\ll \sum_{r|m} r^{-\frac{3}{2}}\phi_{\frac{1}{2}}(r)d(r)\sum_{l=1}^{\infty} l^{-\frac{1}{2}}(l+mr^{-1}K^{-2})^{-1}d(l)$$

$$\ll Km^{-\frac{1}{2}}\log(2m)\sum_{r|m} r^{-1}\phi_{\frac{1}{2}}(r)d(r) \ll d_3(m)Km^{-\frac{1}{2}}\log(2m).$$

Combining these we end the proof of Lemma 2.4.

It should be remarked that the bound (2.3.2) is essentially best possible as far as the dependency on K is concerned. In fact, it is possible to prove an asymptotic formula whose main term is of the order of K^2. Hence, in particular, the cardinality of the eigenvalues of Δ acting over $B^{\infty}(\mathcal{F})$ is infinite, which was already claimed prior to Theorem 1.1. Since this asymptotic formula has no relevance to our purpose, we shall show, to the same effect, the following lower bound: For each fixed $m \geq 1$ we have

$$\sum_{\kappa_j \leq K} |\rho_j(m)|^2 e^{-\pi\kappa_j} \gg K^2 \tag{2.3.14}$$

as K tends to infinity.

To this end we consider the estimation of $P^{(1)}(r,K)$. We have, by the definition (2.3.8),

$$P^{(1)}(r,K) \ll \int_0^{\infty} e^{-\pi|t-r|}\left(e^{-(t/K)^2} - e^{-(2t/K)^2}\right)\frac{dt}{t}$$

uniformly for $r \geq 0$ and $K \geq 1$. If $0 \leq r \leq 1$ then it is easy to see that $P^{(1)}(r,K) \ll 1$. On the other hand, if $r \geq 1$ then we divide the integral into two parts according as $|t-r| \geq \frac{1}{2}r$ or $|t-r| < \frac{1}{2}r$. Obviously the first part contributes $O(e^{-\pi r/2})$. On the other hand the second part is $O(rK^{-2})$ if $r \leq K$, and $O(r^{-1}e^{-(r/(2K))^2})$ otherwise.

Thus the sum over j on the left side of (2.3.7) is

$$\ll \sum_{j=1}^{\infty} |\rho_j(m)|^2 e^{-\frac{3}{2}\pi\kappa_j} + K^{-1}\sum_{\kappa_j \leq K} |\rho_j(m)|^2 e^{-\pi\kappa_j}$$

$$+ \sum_{\kappa_j > K} \kappa_j^{-1}|\rho_j(m)|^2 e^{-\pi\kappa_j-(\kappa_j/(2K))^2},$$

where the implied constant is absolute. The first sum in this converges to a constant because of (2.3.2). The part of the last sum corresponding to $\kappa_j > NK$ is $O(Ne^{-(N/2)^2}K)$. Similarly the integral in (2.3.7) can be estimated to be of the order of a power of $\log K$. Further, we have seen already that the sum on the right side of (2.3.7) is negligible if m is fixed. Collecting these and taking N sufficiently large we readily reach an expression equivalent to (2.3.14).

Now we are ready to treat the spectral sum (2.3.1). We introduce first the condition

$$\left.\begin{aligned} f(r) \text{ is regular in the strip } |\mathrm{Im}\, r| \le \tfrac{1}{2}, \text{ and there} \\ f(r) = f(-r),\ |f(r)| \ll (1+|r|)^{-2-\delta}, \end{aligned}\right\} \tag{2.3.15}$$

where δ is an arbitrary small positive constant (cf. (2.6.24)). The decay property obviously ensures the convergence of the sum (2.3.1) by virtue of (2.3.2).

Theorem 2.2 *Let* $\{\rho_j(n)\}$ *be the Maass–Fourier coefficients defined by* (1.1.41) *and* (1.1.43), *and let* f *satisfy the condition* (2.3.15). *Then we have, for any integers* $m, n \ge 1$,

$$\sum_{j=1}^{\infty} \frac{\overline{\rho_j(m)}\rho_j(n)}{\cosh(\pi\kappa_j)} f(\kappa_j) + \frac{1}{\pi}\int_{-\infty}^{\infty} \frac{\sigma_{2ir}(m)\sigma_{2ir}(n)}{(mn)^{ir}|\zeta(1+2ir)|^2} f(r)dr$$

$$= \frac{1}{\pi^2}\delta_{m,n}\int_{-\infty}^{\infty} r\tanh(\pi r) f(r)dr + \sum_{l=1}^{\infty} \frac{1}{l} S(m,n\,;l) f_+\left(\frac{4\pi}{l}\sqrt{mn}\right), \tag{2.3.16}$$

where

$$f_+(x) = \frac{2i}{\pi}\int_{-\infty}^{\infty} \frac{r}{\cosh(\pi r)} J_{2ir}(x) f(r)dr. \tag{2.3.17}$$

Proof This is another application of the identity (2.3.4): We multiply both sides by the factor

$$\cosh(\pi t) f(t + \tfrac{1}{2}i),$$

and integrate with respect to t over the real axis. On the left side the resulting expression is absolutely convergent, as can be seen from (2.3.2) and

$$\int_{-\infty}^{\infty} |\cosh(\pi t) f(t+\tfrac{1}{2}i) p(t,r)| dt \ll (1+|r|)^{-2-\delta}.$$

Then we are left with the evaluation of

$$\int_{-\infty}^{\infty} \cosh(\pi t) f(t+\tfrac{1}{2}i) p(t,r) dt$$

for real r. Shifting the path to $\operatorname{Im} t = -\frac{1}{2}$ with the indents of small upper semicircles $c^+(\pm r)$ which are centered at $t = \pm r - \frac{1}{2}i$ and of the same diameter, we see that this is equal to

$$i\cosh(\pi r) = \left\{ \int_{c^+(-r)+\frac{1}{2}i} + \int_{c^+(r)+\frac{1}{2}i} \right\} \frac{i\sinh(\pi t)\cosh(\pi r)}{\sinh(\pi(t+r))\sinh(\pi(t-r))} f(t)dt = f(r),$$

where we have used the assumption that f is even. Thus we get the expression on the left side of (2.3.16).

Moving to the other side, we may obviously restrict our consideration to the part containing Kloosterman sums. We have absolute convergence there as well. In fact we see that because of (2.1.4), (2.3.15), and Stirling's formula this part is

$$\ll \sum_{l=1}^{\infty} l^{2\alpha-2}|S(m,n\,;l)|$$
$$\times \int_{-\infty}^{\infty}\int_{-\infty}^{\infty} (1+|t|)^{-1-\delta}(1+|u|)^{-1}((1+|t+u|)(1+|t-u|))^{\alpha-\frac{1}{2}}$$
$$\times \exp\left(-\tfrac{1}{2}\pi(|t+u|+|t-u|-2|u|)\right) dtdu < +\infty,$$

provided $0 < \alpha < \frac{1}{4}$. Thus we are led to the expression

$$\sum_{l=1}^{\infty} \frac{1}{l} S(m,n;l) C\left(\frac{4\pi}{l}\sqrt{mn}\,;f\right), \tag{2.3.18}$$

where

$$C(x;f) = \frac{1}{\pi^2 i}\int_{(\alpha)} \frac{\sin(\pi\eta)}{\pi\eta} D(\eta;f)\left(\frac{x}{2}\right)^{1-2\eta} d\eta$$

with

$$D(\eta;f) = \int_{-\infty}^{\infty} \frac{t}{\tanh(\pi t)} \Gamma(\eta+it)\Gamma(\eta-it) f(t+\tfrac{1}{2}i)dt.$$

The function $D(\eta;f)$ is obviously regular for $\operatorname{Re}\eta > 0$. We then assume temporarily that $\operatorname{Re}\eta > \frac{1}{2}$. We shift the path to $\operatorname{Im} t = -\frac{1}{2}$, getting

$$D(\eta;f) = \int_{-\infty}^{\infty} (t-\tfrac{1}{2}i)\tanh(\pi t)\Gamma(\eta+\tfrac{1}{2}+it)\Gamma(\eta-\tfrac{1}{2}-it)f(t)dt.$$

Using again the assumption that f is even, we have

$$2D(\eta;f) = \int_{-\infty}^{\infty} \tanh(\pi t)\big[(t-\tfrac{1}{2}i)\Gamma(\eta+\tfrac{1}{2}+it)\Gamma(\eta-\tfrac{1}{2}-it)$$
$$+ (t+\tfrac{1}{2}i)\Gamma(\eta+\tfrac{1}{2}-it)\Gamma(\eta-\tfrac{1}{2}+it)\big] f(t)dt.$$

Thus we have

$$D(\eta;f)=\eta\int_{-\infty}^{\infty}t\tanh(\pi t)\Gamma(\eta-\tfrac{1}{2}+it)\Gamma(\eta-\tfrac{1}{2}-it)f(t)dt.$$

The functional equation $\Gamma(s)\Gamma(1-s)=\pi/\sin(\pi s)$ transforms this into

$$\begin{aligned}D(\eta;f)\\&=\frac{\pi i\eta}{2\sin(\pi\eta)}\int_{-\infty}^{\infty}\frac{t}{\cosh(\pi t)}\left\{\frac{\Gamma(\eta-\frac{1}{2}+it)}{\Gamma(\frac{3}{2}-\eta+it)}-\frac{\Gamma(\eta-\frac{1}{2}-it)}{\Gamma(\frac{3}{2}-\eta-it)}\right\}f(t)dt\\&=\frac{\pi i\eta}{\sin(\pi\eta)}\int_{-\infty}^{\infty}\frac{t}{\cosh(\pi t)}\frac{\Gamma(\eta-\frac{1}{2}+it)}{\Gamma(\frac{3}{2}-\eta+it)}f(t)dt.\end{aligned}$$

We shift the path in the last integral to $\operatorname{Im} t=-\beta$ with $0<\beta<\frac{1}{2}$, getting

$$D(\eta;f)=\frac{\pi i\eta}{\sin(\pi\eta)}\int_{\operatorname{Im} t=-\beta}\frac{t}{\cosh(\pi t)}\frac{\Gamma(\eta-\frac{1}{2}+it)}{\Gamma(\frac{3}{2}-\eta+it)}f(t)dt.$$

By analytic continuation this expression holds for $\operatorname{Re}\eta>\frac{1}{2}-\beta$. Then we have

$$C(x;f)=\frac{1}{\pi^2}\int_{(\alpha)}\left(\frac{x}{2}\right)^{1-2\eta}\int_{\operatorname{Im} t=-\beta}\frac{t}{\cosh(\pi t)}\frac{\Gamma(\eta-\frac{1}{2}+it)}{\Gamma(\frac{3}{2}-\eta+it)}f(t)dt\,d\eta,$$

where α is as above and $\alpha+\beta>\frac{1}{2}$. Stirling's formula readily implies the absolute convergence of the double integral. Exchanging the order of integration and using (2.2.17) we see that

$$C(x;f)=\frac{2i}{\pi}\int_{\operatorname{Im} t=-\beta}\frac{t}{\cosh(\pi t)}J_{2it}(x)f(t)dt.$$

Finally we invoke (2.2.8), which allows us to replace the path $\operatorname{Im} t=-\beta$ by the real axis. Inserting the result into (2.3.18) we end the proof of Theorem 2.2.

2.4 Sums of Kloosterman sums

We are now going to reverse the relation (2.3.16); that is, the sum $K_+(m,n;\varphi)$ introduced at the beginning of this chapter is to be expressed in terms of objects related to cusp-forms, where φ is to satisfy certain mild conditions. For this purpose we shall first show that the formula (2.1.2) with $s_1=1$ yields, in a straightforward way, a spectral decomposition of the Kloosterman-sum zeta-function

$$Z_{m,n}(s) = (2\pi\sqrt{mn})^{2s-1} \sum_{l=1}^{\infty} l^{-2s} S(m,n;l) \quad (m,n>0). \tag{2.4.1}$$

Lemma 2.5 *Let $\{\rho_j(n)\}$ be the Maass–Fourier coefficients defined by* (1.1.41) *and* (1.1.43); *and let $q_{m,n}(k)$ be as in* (2.2.6). *Then we have, for any $m,n \geq 1$ and $\operatorname{Re} s > \frac{3}{4}$,*

$$\begin{aligned} Z_{m,n}(s) &= \frac{1}{2}\sin(\pi s)\sum_{j=1}^{\infty} \frac{\overline{\rho_j(m)}\rho_j(n)}{\cosh(\pi\kappa_j)}\Gamma(s-\tfrac{1}{2}+i\kappa_j)\Gamma(s-\tfrac{1}{2}-i\kappa_j) \\ &+ \frac{1}{2\pi}\sin(\pi s)\int_{-\infty}^{\infty} \frac{\sigma_{2ir}(m)\sigma_{2ir}(n)}{(mn)^{ir}|\zeta(1+2ir)|^2}\Gamma(s-\tfrac{1}{2}+ir)\Gamma(s-\tfrac{1}{2}-ir)dr \\ &+ \sum_{k=1}^{\infty}(2k-1)q_{m,n}(k)\frac{\Gamma(k-1+s)}{\Gamma(k+1-s)} - \frac{1}{2\pi}\delta_{m,n}\frac{\Gamma(s)}{\Gamma(1-s)}. \end{aligned} \tag{2.4.2}$$

Proof We specialize (2.1.2) by putting

$$s_1 = 1, \quad s_2 = s$$

with

$$\operatorname{Re} s > 1-\alpha, \quad 0<\alpha<\tfrac{1}{4}.$$

We note that

$$W(x\,;1,s) = \frac{1}{2\pi i}\int_{(\alpha)} \frac{\Gamma(\eta)}{\Gamma(1-\eta)(\eta+s-1)}\left(\frac{x}{2}\right)^{-2\eta} d\eta.$$

We are going to transform this into a series of the Neumann type: The elementary identity

$$\begin{aligned} &\frac{\Gamma(k+s)\Gamma(k+\eta)}{\Gamma(k+1-s)\Gamma(k+1-\eta)} - \frac{\Gamma(k-1+s)\Gamma(k-1+\eta)}{\Gamma(k-s)\Gamma(k-\eta)} \\ &\qquad = (\eta+s-1)(2k-1)\frac{\Gamma(k-1+s)\Gamma(k-1+\eta)}{\Gamma(k+1-s)\Gamma(k+1-\eta)} \end{aligned}$$

gives

$$\begin{aligned} \frac{\Gamma(s)\Gamma(\eta)}{\Gamma(1-s)\Gamma(1-\eta)(\eta+s-1)} &= \frac{\Gamma(K+s)\Gamma(K+\eta)}{\Gamma(K+1-s)\Gamma(K+1-\eta)(\eta+s-1)} \\ &\quad - \sum_{k=1}^{K}(2k-1)\frac{\Gamma(k-1+s)\Gamma(k-1+\eta)}{\Gamma(k+1-s)\Gamma(k+1-\eta)}. \end{aligned}$$

This and (2.2.17) imply that

$$\frac{\Gamma(s)}{\Gamma(1-s)}W(x;1,s)$$
$$= \frac{1}{2\pi i}\int_{(\alpha)} \frac{\Gamma(K+s)\Gamma(K+\eta)}{\Gamma(K+1-s)\Gamma(K+1-\eta)(\eta+s-1)}\left(\frac{x}{2}\right)^{-2\eta} d\eta$$
$$- \frac{2}{x}\sum_{k=1}^{K}(2k-1)\frac{\Gamma(k-1+s)}{\Gamma(k+1-s)}J_{2k-1}(x).$$

As K tends to infinity the integral converges to $2\pi i(x/2)^{2(s-1)}$, which comes from the simple pole at $\eta = 1-s$. To see this we move the path to $\operatorname{Re}\eta = -\beta$ with a large positive $\beta < K$, passing over $1-s$, and note that the resulting integrand is, by Stirling's formula,

$$\ll K^{2\operatorname{Re}s-1}(K+|\eta|)^{-2\beta-1},$$

which gives the assertion. Hence we find that

$$\frac{\Gamma(s)}{\Gamma(1-s)}W(x;1,s) = \left(\frac{x}{2}\right)^{2(s-1)} - 2x^{-1}\sum_{k=1}^{\infty}(2k-1)\frac{\Gamma(k-1+s)}{\Gamma(k+1-s)}J_{2k-1}(x). \tag{2.4.3}$$

In this we put $x = 4\pi\sqrt{mn}/l$, and multiply both sides of the resulting identity by $2\pi\sqrt{mn}S(m,n;l)l^{-2}$. Then we sum over l, getting

$$Z_{m,n}(s) = 2\pi\sqrt{mn}\frac{\Gamma(s)}{\Gamma(1-s)}\sum_{l=1}^{\infty} l^{-2}S(m,n;l)W\left(\frac{4\pi}{l}\sqrt{mn};1,s\right)$$
$$+\sum_{l=1}^{\infty} S(m,n;l)l^{-1}\sum_{k=1}^{\infty}(2k-1)\frac{\Gamma(k-1+s)}{\Gamma(k+1-s)}J_{2k-1}\left(\frac{4\pi}{l}\sqrt{mn}\right).$$

The bound (2.2.7) implies the double sum is absolutely convergent; thus by (2.1.2) and (2.2.6) we have

$$Z_{m,n}(s) = 2\sqrt{mn}(4\pi n)^{s-1}\langle P_m(\cdot,1), P_n(\cdot,\bar{s})\rangle/\Gamma(1-s)$$
$$+\sum_{k=1}^{\infty}(2k-1)q_{m,n}(k)\frac{\Gamma(k-1+s)}{\Gamma(k+1-s)} - \frac{1}{2\pi}\delta_{m,n}\frac{\Gamma(s)}{\Gamma(1-s)}.$$

Into this we insert (2.1.14), and end the proof of the lemma.

Having established (2.4.2), we are ready to prove the spectral decomposition of $K_+(m,n;\varphi)$. The relation between $Z_{m,n}(s)$ and $K_+(m,n;\varphi)$ is immediate: We have

$$K_+(m,n;\varphi) = \frac{1}{2\pi i}\int_{(\alpha)} Z_{m,n}(s)\varphi^*(s)ds, \tag{2.4.4}$$

where

$$\varphi^*(s) = \int_0^\infty \varphi(x)\left(\frac{x}{2}\right)^{-2s} dx, \tag{2.4.5}$$

and α is to be chosen appropriately. We are going to insert the expression (2.4.2) into (2.4.4). Then we shall encounter a convergence problem. To overcome it we have to impose some conditions on $\varphi(x)$. To this end we assume that $\varphi(x) \in C^3(0,\infty)$ and for $\nu = 0, 1, 2, 3$

$$\varphi^{(\nu)}(x) \ll \begin{cases} x^{\frac{1}{2}+\delta-\nu} & \text{as } x \to +0, \\ \\ x^{-1-\delta-\nu} & \text{as } x \to +\infty, \end{cases} \tag{2.4.6}$$

where δ is an arbitrary small positive constant. This set of conditions is a result of compromise; so there are possibilities of relaxing them.

Theorem 2.3 *Let $\{\rho_j(n)\}$ be the Maass–Fourier coefficients defined by* (1.1.41) *and* (1.1.43); *and let $\{\rho_{j,k}(n)\}$ be as in* (2.2.2)–(2.2.3). *Further, let φ satisfy the condition* (2.4.6). *Then we have, for any integers $m, n \geq 1$,*

$$\begin{aligned} K_+(m, n\,; \varphi) = &\sum_{j=1}^\infty \frac{\overline{\rho_j(m)}\rho_j(n)}{\cosh(\pi\kappa_j)}\varphi^+(\kappa_j) \\ &+ \frac{1}{\pi}\int_{-\infty}^\infty \frac{\sigma_{2ir}(m)\sigma_{2ir}(n)}{(mn)^{ir}|\zeta(1+2ir)|^2}\varphi^+(r)dr \\ &+ 2\sum_{k=1}^\infty a(k)\sum_{j=1}^{\vartheta(k)} \overline{\rho_{j,k}(m)}\rho_{j,k}(n)\varphi^+((\tfrac{1}{2}-k)i), \end{aligned} \tag{2.4.7}$$

where

$$\varphi^+(r) = \frac{\pi i}{2\sinh(\pi r)}\int_0^\infty (J_{2ir}(x) - J_{-2ir}(x))\varphi(x)\frac{dx}{x}. \tag{2.4.8}$$

Proof Providing (2.4.6) holds, the transform $\varphi^*(s)$ is regular in the strip

$$-\tfrac{1}{2}\delta < \operatorname{Re} s < \tfrac{3}{4} + \tfrac{1}{2}\delta. \tag{2.4.9}$$

Integrating three times by parts in (2.4.5) we have, for s in (2.4.9),

$$\varphi^*(s) \ll (1 + |s|)^{-3}. \tag{2.4.10}$$

This gives, in particular, the Mellin inversion

$$\varphi(x) = \frac{1}{2\pi i}\int_{(\alpha)} \varphi^*(s)\left(\frac{x}{2}\right)^{2s-1} ds \tag{2.4.11}$$

with any line $\operatorname{Re} s = \alpha$ in the strip (2.4.9). Hence the expression (2.4.4) holds if

$$\tfrac{3}{4} < \alpha < \tfrac{3}{4} + \tfrac{1}{2}\delta. \tag{2.4.12}$$

Supposing this we insert (2.4.2) into (2.4.4), and have

$$\begin{aligned} K_{+}(m,n\,;\varphi) &= \sum_{j=1}^{\infty} \frac{\overline{\rho_j(m)}\rho_j(n)}{\cosh(\pi\kappa_j)} \frac{1}{2\pi i}\int_{(\alpha)} \varphi^{*}(s)h(s,\kappa_j)ds \\ &+ \frac{1}{\pi}\int_{-\infty}^{\infty} \frac{\sigma_{2ir}(m)\sigma_{2ir}(n)}{(mn)^{ir}|\zeta(1+2ir)|^2} \frac{1}{2\pi i}\int_{(\alpha)} \varphi^{*}(s)h(s,r)dsdr \\ &+ \sum_{k=1}^{\infty}(2k-1)q_{m,n}(k)\frac{1}{2\pi i}\int_{(\alpha)} \varphi^{*}(s)\frac{\Gamma(k-1+s)}{\Gamma(k+1-s)}ds \\ &- \frac{1}{2\pi}\delta_{m,n}\cdot\frac{1}{2\pi i}\int_{(\alpha)} \varphi^{*}(s)\frac{\Gamma(s)}{\Gamma(1-s)}ds, \end{aligned} \tag{2.4.13}$$

where

$$h(s,r) = \frac{1}{2}\sin(\pi s)\Gamma(s-\tfrac{1}{2}+ir)\Gamma(s-\tfrac{1}{2}-ir).$$

This has to be verified. So we consider the expression

$$\sum_{j=1}^{\infty} |\rho_j(m)\rho_j(n)|e^{-\pi\kappa_j}\int_{(\alpha)} |\varphi^{*}(s)h(s,\kappa_j)||ds|. \tag{2.4.14}$$

By (2.4.10) and Stirling's formula we have, for real r and $s = \alpha + it$,

$$\begin{aligned} \varphi^{*}(s)h(s,r) \ll \exp\big(-\tfrac{1}{2}\pi(|t+r|+|t-r|-2|t|)\big) \\ \times ((1+|t+r|)(1+|t-r|))^{\alpha-1}(1+|t|)^{-3}, \end{aligned}$$

and thus

$$\int_{(\alpha)} |\varphi^{*}(s)h(s,r)||ds| \ll (1+|r|)^{2\alpha-4}.$$

Then by (2.3.2) we see that (2.4.14) is finite, given the condition (2.4.12). The last estimate implies also that the double integral on the right side of (2.4.13) is absolutely convergent. Further, we easily have

$$\int_{(\alpha)} \left|\varphi^{*}(s)\frac{\Gamma(k-1+s)}{\Gamma(k+1-s)}\right||ds| \ll k^{2(\alpha-1)},$$

which with (2.2.6) and (2.2.7) gives the absolute convergence of the sum over k in (2.4.13). Hence we have confirmed (2.4.13).

We shall show that integrals over s in (2.4.13) can be expressed in terms of φ and J-Bessel functions. But before doing this we rearrange

(2.4.13) a little: We note that for any integer $K \geq 1$

$$\frac{\Gamma(s)}{\Gamma(1-s)} = \sum_{k=1}^{K}(-1)^{k-1}\left\{\frac{\Gamma(k+s)}{\Gamma(k+1-s)} + \frac{\Gamma(k-1+s)}{\Gamma(k-s)}\right\} + (-1)^K\frac{\Gamma(K+s)}{\Gamma(K+1-s)}$$
$$= \sum_{k=1}^{K}(-1)^{k-1}(2k-1)\frac{\Gamma(k+s)}{\Gamma(k+1-s)} + (-1)^K\frac{\Gamma(K+s)}{\Gamma(K+1-s)}.$$

We multiply this identity by $\varphi^*(s)$ and integrate over the line $\operatorname{Re} s = \alpha$. In the last integral of the resulting identity we shift the contour to $\operatorname{Re} s = 0$, and see that it tends to 0 as K goes to $+\infty$. Thus we have, given (2.4.12),

$$\int_{(\alpha)} \varphi^*(s)\frac{\Gamma(s)}{\Gamma(1-s)}ds = \sum_{k=1}^{\infty}(-1)^{k-1}(2k-1)\int_{(\alpha)} \varphi^*(s)\frac{\Gamma(k-1+s)}{\Gamma(k+1-s)}ds.$$

This and (2.2.9) imply that the sum of the last two terms on the right of (2.4.13) is equal to

$$\pi\sum_{k=1}^{\infty}(-1)^k a(k)\sum_{j=1}^{\vartheta(k)}\overline{\rho_{j,k}(m)}\rho_{j,k}(n)\frac{1}{2\pi i}\int_{(\alpha)}\varphi^*(s)\frac{\Gamma(k-1+s)}{\Gamma(k+1-s)}ds. \tag{2.4.15}$$

Also we note that

$$\frac{1}{2\pi i}\int_{(\alpha)}\varphi^*(s)h(s,r)ds = u(r,\varphi) + u(-r,\varphi), \tag{2.4.16}$$

where

$$u(r,\varphi) = \frac{1}{8\sinh(\pi r)}\int_{(\alpha)}\varphi^*(s)\frac{\Gamma(s-\frac{1}{2}+ir)}{\Gamma(\frac{3}{2}-s+ir)}ds. \tag{2.4.17}$$

From (2.4.13), (2.4.15), and (2.4.16) we get

$$K_+(m,n;\varphi) = \sum_{j=1}^{\infty}\frac{\overline{\rho_j(m)}\rho_j(n)}{\cosh(\pi\kappa_j)}(u(\kappa_j,\varphi) + u(-\kappa_j,\varphi))$$
$$+ \frac{2}{\pi}\int_{-\infty}^{\infty}\frac{\sigma_{2ir}(m)\sigma_{2ir}(n)}{(mn)^{ir}|\zeta(1+2ir)|^2}u(r,\varphi)dr$$
$$+ 4\sum_{k=1}^{\infty}a(k)\sum_{j=1}^{\vartheta(k)}\overline{\rho_{j,k}(m)}\rho_{j,k}(n)u((\tfrac{1}{2}-k)i,\varphi). \tag{2.4.18}$$

Now we express $u(r,\varphi)$ in terms of φ. We treat only the case where r is real, for $u((\frac{1}{2}-k)i,\varphi)$ can be dealt with in much the same way. We observe first that the integral in (2.4.17) is regular for $\operatorname{Im} r < \alpha - \frac{1}{2}$. Let us suppose temporarily that $\operatorname{Im} r = -\frac{1}{4}$, say. Then we may move the path to $\operatorname{Re} s = \frac{1}{3}$. Inserting (2.4.5) into the resulting integral, we get an absolutely convergent double integral. After exchanging the order of integration we have

$$\int_{(\frac{1}{3})} \cdots ds = \int_0^\infty \varphi(x) \int_{(\frac{1}{3})} \frac{\Gamma(s - \frac{1}{2} + ir)}{\Gamma(\frac{3}{2} - s + ir)} \left(\frac{x}{2}\right)^{-2s} ds dx$$
$$= 4\pi i \int_0^\infty J_{2ir}(x)\varphi(x)\frac{dx}{x},$$

where we have used (2.2.17). But the last integral is regular in the strip $-\frac{1}{2} < \operatorname{Im} r < \frac{1}{4}$; this is an easy consequence of the combination of (2.4.6) and the representation (2.2.8). Hence, by analytic continuation, we have, for real r,

$$u(r, \varphi) = \frac{\pi i}{2\sinh(\pi r)} \int_0^\infty J_{2ir}(x)\varphi(x)\frac{dx}{x}. \tag{2.4.19}$$

It can also be shown that this identity holds for $r = (\frac{1}{2} - k)i$ with integral $k \geq 1$. Combining (2.4.18) and (2.4.19) we end the proof of Theorem 2.3.

2.5 Opposite-sign case

Next we shall prove the counterparts of (2.3.16) and (2.4.7) for the situation involving the Kloosterman sums $S(m, -n; l)$ $(m, n \geq 1)$. But we shall be brief, for our new problem is far simpler than the same-sign case which has been developed above.

This time we consider the inner-product $\langle P_m(\cdot, s_1), \overline{P_n(\cdot, s_2)}\rangle$. The procedure used to derive (2.1.2) readily gives

$$\langle P_m(\cdot, s_1), \overline{P_n(\cdot, s_2)}\rangle = 2^{2(1-s_2)}\pi^{s_1-s_2+1}n^{s_1-s_2}\frac{\Gamma(s_1 + s_2 - 1)}{\Gamma(s_1)\Gamma(s_2)}$$
$$\times \sum_{l=1}^\infty l^{-2s_1} S(m, -n\,; l) W_-\left(\frac{4\pi}{l}\sqrt{mn}; s_1, s_2\right), \tag{2.5.1}$$

where s_1, s_2, α are to satisfy (2.1.1), and

$$W_-(x; s_1, s_2) = \frac{1}{2\pi i} \int_{(\alpha)} \Gamma(\eta)\Gamma(\eta + s_2 - s_1)\left(\frac{x}{2}\right)^{-2\eta} d\eta.$$

We note that W_- is essentially a K-Bessel function, for we have

$$K_{2\nu}(x) = \frac{1}{4\pi i} \int_{(\beta)} \Gamma(s + \nu)\Gamma(s - \nu)\left(\frac{x}{2}\right)^{-2s} ds, \tag{2.5.2}$$

provided $\beta > |\operatorname{Re} \nu|$, which can be confirmed easily by expressing the exponentiated factor of (1.1.17) in terms of a Mellin inversion. We have also the spectral decomposition

$$\langle P_m(\cdot,s_1),\overline{P_n(\cdot,s_2)}\rangle = \frac{\pi}{\Gamma(s_1)\Gamma(s_2)}(4\pi\sqrt{mn})^{1-s_1-s_2}(n/m)^{\frac{1}{2}(s_1-s_2)}$$
$$\times\Bigg\{\sum_{j=1}^{\infty}\overline{\rho_j(m)}\rho_j(-n)\Theta(s_1,s_2;\kappa_j)$$
$$+\frac{1}{\pi}\int_{-\infty}^{\infty}\frac{\sigma_{2ir}(m)\sigma_{2ir}(n)\cosh(\pi r)\Theta(s_1,s_2;r)}{(mn)^{ir}|\zeta(1+2ir)|^2}dr\Bigg\}, \tag{2.5.3}$$

where $\mathrm{Re}\, s_1, \mathrm{Re}\, s_2 > \frac{3}{4}$, and Θ is defined by (2.1.15).

Then we specialize (2.5.1) and (2.5.3) by (2.3.3), getting

$$\sum_{j=1}^{\infty}\frac{\overline{\rho_j(m)}\rho_j(-n)}{\cosh(\pi\kappa_j)}p(t,\kappa_j)+\frac{1}{\pi}\int_{-\infty}^{\infty}\frac{\sigma_{2ir}(m)\sigma_{2ir}(n)}{(mn)^{ir}|\zeta(1+2ir)|^2}p(t,r)dr$$
$$=\frac{2}{\pi^2}\sum_{l=1}^{\infty}\frac{1}{l}S(m,-n\,;l)\varpi_-\Big(t,\frac{4\pi}{l}\sqrt{mn}\Big), \tag{2.5.4}$$

where $p(t,r)$ is as in (2.3.4), and

$$\varpi_-(t,x)=\frac{1}{2\pi i}\int_{(\alpha)}\Gamma(\eta+it)\Gamma(\eta-it)\Big(\frac{x}{2}\Big)^{1-2\eta}d\eta$$

with t, α satisfying (2.3.6). As before we let f satisfy (2.3.15), multiply both sides of (2.5.4) by $\cosh(\pi t)f(t+\frac{1}{2}i)$, and integrate with respect to t over the real axis; here the absolute convergence is easy to check. Then we are left with the evaluation of

$$\int_{-\infty}^{\infty}\varpi_-(t,x)\cosh(\pi t)f(t+\tfrac{1}{2}i)dt.$$

This is equal to

$$\frac{1}{2\pi i}\int_{(\alpha)}\Big(\frac{x}{2}\Big)^{1-2\eta}\int_{-\infty}^{\infty}\cosh(\pi t)\Gamma(\eta+it)\Gamma(\eta-it)f(t+\tfrac{1}{2}i)dtd\eta$$
$$=\frac{1}{2\pi i}\int_{(\beta)}\Big(\frac{x}{2}\Big)^{1-2\eta}\int_{-\infty}^{\infty}\cdots dtd\eta \tag{2.5.5}$$

with $\beta=\frac{1}{2}(1+\frac{1}{2}\delta)$. We shift the contour in the last inner integral to

$\operatorname{Im} t = -\frac{1}{2}$; then it becomes

$$
\begin{aligned}
&-i\int_{-\infty}^{\infty} \sinh(\pi t)\Gamma(\eta+it+\tfrac{1}{2})\Gamma(\eta-it-\tfrac{1}{2})f(t)dt \\
&= -\frac{i}{2}\int_{-\infty}^{\infty} \sinh(\pi t)\left\{\Gamma(\eta+it+\tfrac{1}{2})\Gamma(\eta-it-\tfrac{1}{2})\right. \\
&\qquad\qquad \left. -\Gamma(\eta-it+\tfrac{1}{2})\Gamma(\eta+it-\tfrac{1}{2})\right\} f(t)dt \\
&= \int_{-\infty}^{\infty} t\sinh(\pi t)\Gamma(\eta+it-\tfrac{1}{2})\Gamma(\eta-it-\tfrac{1}{2})f(t)dt.
\end{aligned}
$$

We insert this into (2.5.5), exchange the order of integration, and apply (2.5.2) to the resulting inner integral.

In this way we obtain

Theorem 2.4 *Let* $\{\rho_j(n)\}$ *be the Maass–Fourier coefficients defined by* (1.1.41) *and* (1.1.43); *and let* f *satisfy* (2.3.15). *Then we have, for any integers* $m, n \geq 1$,

$$
\begin{aligned}
&\sum_{j=1}^{\infty} \frac{\overline{\rho_j(m)}\rho_j(-n)}{\cosh(\pi\kappa_j)} f(\kappa_j) + \frac{1}{\pi}\int_{-\infty}^{\infty} \frac{\sigma_{2ir}(m)\sigma_{2ir}(n)}{(mn)^{ir}|\zeta(1+2ir)|^2} f(r)dr \\
&\qquad = \sum_{l=1}^{\infty} \frac{1}{l} S(m,-n\,;l) f_-\left(\frac{4\pi}{l}\sqrt{mn}\right),
\end{aligned} \tag{2.5.6}
$$

where

$$
f_-(x) = \frac{4}{\pi^2}\int_{-\infty}^{\infty} r\sinh(\pi r)K_{2ir}(x)f(r)dr. \tag{2.5.7}
$$

We now turn to $K_-(m,n;\varphi)$ introduced at the beginning of this chapter, and assume that φ satisfies the condition (2.4.6). Unfortunately we do not have an analogue of (2.4.2) for our present situation, which would make our problem easier. We can, however, come close to it. This is by virtue of the fact that

$$
W_-(x\,;s,s+\tfrac{1}{2}) = \sqrt{\pi}e^{-x}, \tag{2.5.8}
$$

which is equivalent to the identity $K_{\frac{1}{2}}(x) = (\pi/2x)^{\frac{1}{2}}e^{-x}$. The specialization of (2.5.1) and (2.5.3) that is indicated by (2.5.8) gives, for $\operatorname{Re} s > \frac{3}{4}$,

$$
\begin{aligned}
&(2\pi\sqrt{mn})^{2s-1}\sum_{l=1}^{\infty} \frac{1}{l^{2s}} S(m,-n\,;l)\exp\left(-\frac{4\pi}{l}\sqrt{mn}\right) \\
&= \sum_{j=1}^{\infty} \frac{\overline{\rho_j(m)}\rho_j(-n)}{\cosh(\pi\kappa_j)}\lambda(s,\kappa_j) + \frac{1}{\pi}\int_{-\infty}^{\infty} \frac{\sigma_{2ir}(m)\sigma_{2ir}(n)}{(mn)^{ir}|\zeta(1+2ir)|^2}\lambda(s,r)dr,
\end{aligned} \tag{2.5.9}
$$

where

$$\lambda(s,r) = \frac{1}{2\sqrt{\pi}} \cosh(\pi r) \frac{\Theta(s, s+\frac{1}{2}; r)}{\Gamma(2s-\frac{1}{2})}$$
$$= 2^{3-4s}\sqrt{\pi}\cosh(\pi r)\frac{\Gamma(2s-1+2ir)\Gamma(2s-1-2ir)}{\Gamma(2s-\frac{1}{2})}.$$

We note that we have, by (2.1.21),

$$\lambda(s,r) = 2\cosh(\pi r)\int_0^\infty e^{-x}K_{2ir}(x)\left(\frac{x}{2}\right)^{2s-1}\frac{dx}{x}. \tag{2.5.10}$$

We are going to extract the sum $K_-(m,n;\varphi)$ from (2.5.9) by integrating the latter. But the presence of the exponential function on the left side of (2.5.9) causes minor trouble. To overcome this we introduce a C^∞ function ω_X depending on a positive parameter X such that

$$\omega_X(x) = \begin{cases} 1 & \text{for } x \le X, \\ \\ 0 & \text{for } x \ge 2X; \end{cases} \tag{2.5.11}$$

and for each fixed $\nu \ge 0$

$$\omega_X^{(\nu)}(x) \ll X^{-\nu}. \tag{2.5.12}$$

We then put

$$\theta_X(x) = e^x\omega_X(x)\varphi(x),$$

and also

$$\theta_X^*(s) = \int_0^\infty \theta_X(x)\left(\frac{x}{2}\right)^{-2s}dx.$$

We obviously have

$$K_-(m,n\,;\varphi) = \sum_{l=1}^\infty \frac{1}{l}S(m,-n\,;l)\exp\left(-\frac{4\pi}{l}\sqrt{mn}\right)\theta_X\left(\frac{4\pi}{l}\sqrt{mn}\right)$$

whenever $X > 4\pi\sqrt{mn}$ which we suppose hereafter. We follow closely the argument that we used to prove (2.4.13). Then we see that the formula (2.5.9) gives

$$K_-(m,n;\varphi) = \sum_{j=1}^\infty \frac{\overline{\rho_j(m)}\rho_j(-n)}{\cosh(\pi\kappa_j)}\lambda_X(\kappa_j)$$
$$+\frac{1}{\pi}\int_{-\infty}^\infty \frac{\sigma_{2ir}(m)\sigma_{2ir}(n)}{(mn)^{ir}|\zeta(1+2ir)|^2}\lambda_X(r)dr, \tag{2.5.13}$$

where

$$\lambda_X(r) = \frac{1}{2\pi i}\int_{(\alpha)} \lambda(s,r)\theta_X^*(s)ds$$

with α as in (2.4.12). But (2.5.10) obviously gives

$$\lambda_X(r) = 2\cosh(\pi r)\int_0^\infty \omega_X(x)\varphi(x)K_{2ir}(x)\frac{dx}{x}.$$

So we have, by (2.5.2),

$$\begin{aligned}
&\varphi^-(r) - \lambda_X(r)\\
&= 2\cosh(\pi r)\int_0^\infty (1-\omega_X(x))\varphi(x)K_{2ir}(x)\frac{dx}{x}\\
&= \frac{1}{4\pi i}\cosh(\pi r)\int_{(\beta)} \Gamma(s-\tfrac{1}{2}+ir)\Gamma(s-\tfrac{1}{2}-ir)((1-\omega_X)\varphi)^*(s)ds,
\end{aligned}$$

where φ^- is defined by (2.5.15) below, $\beta > \frac{1}{2}$, and

$$((1-\omega_X)\varphi)^*(s) = \int_0^\infty (1-\omega_X(x))\varphi(x)\left(\frac{x}{2}\right)^{-2s} dx.$$

The suppositions (2.4.6) and (2.5.12) imply that $((1-\omega_X)\varphi)^*(s)$ is regular for $\operatorname{Re} s \ge -\frac{1}{4}\delta$, and there

$$((1-\omega_X)\varphi)^*(s) \ll X^{-\frac{1}{2}\delta}(|s|+1)^{-3}.$$

Thus we have

$$\begin{aligned}
\varphi^-(r) - \lambda_X(r) = {}& \frac{1}{4\pi i}\cosh(\pi r)\int_{(-\frac{1}{4}\delta)} \cdots\, ds\\
&+ \frac{1}{2}\cosh(\pi r)\left\{((1-\omega_X)\varphi)^*(\tfrac{1}{2}+ir)\Gamma(2ir)\right.\\
&\qquad\left. +((1-\omega_X)\varphi)^*(\tfrac{1}{2}-ir)\Gamma(-2ir)\right\},
\end{aligned}$$

which is $O(X^{-\frac{1}{2}\delta}(1+|r|)^{-2-\frac{1}{2}\delta})$. We insert this into (2.5.13) and let X tend to infinity, while invoking (2.3.2).

In this way we obtain

Theorem 2.5 *Let $\{\rho_j(n)\}$ be the Maass–Fourier coefficients defined by* (1.1.41) *and* (1.1.43); *and let φ satisfy the condition* (2.4.6). *Then we have, for any integers $m, n \geq 1$,*

$$\sum_{l=1}^{\infty} \frac{1}{l} S(m, -n\,; l)\varphi\left(\frac{4\pi}{l}\sqrt{mn}\right)$$
$$= \sum_{j=1}^{\infty} \frac{\overline{\rho_j(m)}\rho_j(-n)}{\cosh(\pi\kappa_j)}\varphi^-(\kappa_j) + \frac{1}{\pi}\int_{-\infty}^{\infty} \frac{\sigma_{2ir}(m)\sigma_{2ir}(n)}{(mn)^{ir}|\zeta(1+2ir)|^2}\varphi^-(r)dr, \tag{2.5.14}$$

where

$$\varphi^-(r) = 2\cosh(\pi r)\int_0^{\infty} \varphi(x)K_{2ir}(x)\frac{dx}{x}. \tag{2.5.15}$$

2.6 Alternative argument

Although the above is quite adequate for our later purpose, we are now going to show different proofs of the trace formulas (2.3.16) and (2.5.6). This will be achieved by appealing to a pair of integral formulas of Barnes' type (Lemma 2.7 below), which in turn depend on an inversion formula for K-Bessel transforms. Because of this we begin with a proof of the latter. What follows is, however, mainly of theoretical interest; thus the precise discussion on the issues like convergence conditions will be left for readers.

Lemma 2.6 *Let us suppose that f is sufficiently smooth on the positive real axis and decays rapidly at both ends of the line. Then we have*

$$f(y) = \pi^{-2}\int_{-\infty}^{\infty} r\sinh(\pi r)K_{ir}(y)\int_0^{\infty} f(v)K_{ir}(v)v^{-1}dv dr. \tag{2.6.1}$$

Proof First we modify the assertion (1.1.22): Let the differential operator

$$D_s = -y^2(d/dy)^2 + s(s-1) + y^2$$

be defined over $C^2(0,\infty)$, and let $g_s(y,v)$ be the resolvent kernel of $y^{-2}D_s$. A simple change of variable in Lemma 1.3 gives, for $y, v > 0$,

$$g_s(y,v) = \begin{cases} \sqrt{yv}\,I_{s-\frac{1}{2}}(v)K_{s-\frac{1}{2}}(y) & \text{if } v \leq y, \\ \sqrt{yv}\,I_{s-\frac{1}{2}}(y)K_{s-\frac{1}{2}}(v) & \text{if } v \geq y. \end{cases} \tag{2.6.2}$$

We shall show that we have alternatively

$$g_s(y,v) = \pi^{-2}\sqrt{yv}\int_{-\infty}^{\infty} \frac{r}{(s-\frac{1}{2})^2+r^2}\sinh(\pi r)K_{ir}(y)K_{ir}(v)dr, \qquad (2.6.3)$$

provided $\mathrm{Re}\, s > \frac{1}{2}$. To this end we denote the right side by $\tilde{g}_s(y,v)$. Then we have, by (1.1.25),

$$\tilde{g}_s(y,v) = i\pi^{-1}\sqrt{yv}\int_{-\infty}^{\infty} \frac{r}{(s-\frac{1}{2})^2+r^2} I_{ir}(y)K_{ir}(v)dr.$$

We shift the path to $\mathrm{Im}\, r = -N$ with a large $N > 0$, passing over the simple pole at $r = i(\frac{1}{2}-s)$. On the new path we put $r = t - iN$, $t \in \mathbb{R}$, and note that we have, by (1.1.24)–(1.1.25),

$$I_{ir}(y)K_{ir}(v) \ll \frac{1}{N+|t|}(y/v)^N$$

with the implied constant being uniform in t, N. Hence, computing the residue and taking N to infinity, we find that $\tilde{g}_s(y,v) = g_s(y,v)$ provided $v \geq y$. This yields (2.6.3) because of symmetry.

We move to the proof of (2.6.1). We assume first that f has a compact support on the positive real axis. We naturally have, for any $y > 0$,

$$D_s \int_0^{\infty} g_s(y,v)f(v)dv = y^2 f(y). \qquad (2.6.4)$$

We replace $g_s(y,v)$ by the expression (2.6.3), getting a double integral which obviously converges absolutely. We exchange the order of integration, and to secure the relevant uniformity of convergence we need a good estimate of the resulting inner integral. Combining (2.5.2) and a simple estimate of the Mellin transform of f based on the condition imposed on f, we can show that the integral is $O((1+|r|)^{-3}\exp(-\frac{1}{2}\pi|r|))$. Having this, we may shift D_s inside the outer integral. On noting that

$$D_s\{\sqrt{y}K_{ir}(y)\} = \left((s-\tfrac{1}{2})^2 + r^2\right)\sqrt{y}K_{ir}(y),$$

we find that (2.6.4) is equivalent to the identity

$$y^2 f(y) = \pi^{-2}\int_{-\infty}^{\infty} r\sinh(\pi r)\sqrt{y}K_{ir}(y)\int_0^{\infty} f(v)\sqrt{v}K_{ir}(v)dvdr.$$

Replacing $f(y)$ by $f(y)y^{-\frac{3}{2}}$, we obtain (2.6.1) for f satisfying the present restriction. In the general case we replace f by $\omega_X f$, and proceed as in the proof of the last theorem. This ends the proof of the lemma.

We are now ready to prove

Lemma 2.7 *We have*

$$\frac{i}{\pi}\int_{-i\infty}^{i\infty}\xi\sin(\pi\xi)\Gamma(\omega_1+\xi)\Gamma(\omega_2+\xi)\Gamma(\omega_3+\xi)$$
$$\times\Gamma(\omega_1-\xi)\Gamma(\omega_2-\xi)\Gamma(\omega_3-\xi)d\xi$$
$$=\Gamma(\omega_1+\tfrac{1}{2})\Gamma(\omega_2+\tfrac{1}{2})\Gamma(\omega_3+\tfrac{1}{2})\Gamma(\omega_1+\omega_2)$$
$$\times\Gamma(\omega_1+\omega_3)\Gamma(\omega_2+\omega_3)/\Gamma(\omega_1+\omega_2+\omega_3+\tfrac{1}{2}) \quad (2.6.5)$$

and

$$\frac{i}{2\pi^2}\int_{-i\infty}^{i\infty}\xi\sin(2\pi\xi)\Gamma(\omega_1+\xi)\Gamma(\omega_2+\xi)\Gamma(\omega_3+\xi)\Gamma(\omega_1-\xi)\Gamma(\omega_2-\xi)$$
$$\times\Gamma(\omega_3-\xi)d\xi=\Gamma(\omega_1+\omega_2)\Gamma(\omega_1+\omega_3)\Gamma(\omega_2+\omega_3). \quad (2.6.6)$$

Here the paths in both integrals are the same. This path is curved to ensure that the poles of $\Gamma(\omega_1+\xi)\Gamma(\omega_2+\xi)\Gamma(\omega_3+\xi)$ *lie to the left of it, and those of* $\Gamma(\omega_1-\xi)\Gamma(\omega_2-\xi)\Gamma(\omega_3-\xi)$ *to the right. It is assumed that parameters* $\omega_1,\omega_2,\omega_3$ *are such that the path can be drawn.*

Proof We consider (2.6.5) first. We start with the formula

$$\int_0^\infty \exp\left(-x-\rho^2/(8x)\right)K_\nu(x)x^{-\frac{3}{2}}dx=2^{\frac{5}{2}}\sqrt{\pi}\rho^{-1}K_{2\nu}(\rho)\quad(\rho>0). \quad (2.6.7)$$

To show this we use (1.1.17). The left side becomes a double integral which is absolutely convergent. Exchanging the order of integration we get the expression

$$\frac{1}{2}\int_0^\infty y^{\nu-1}\int_0^\infty x^{-\frac{3}{2}}\exp\Big(-\tfrac{1}{2}x(\sqrt{y}+1/\sqrt{y})^2-\rho^2/(8x)\Big)dxdy,$$

which is equal to

$$(2\rho^{-1})^{\frac{1}{2}}\int_0^\infty y^{\nu-1}(\sqrt{y}+1/\sqrt{y})^{\frac{1}{2}}K_{\frac{1}{2}}\Big(\tfrac{1}{2}\rho(\sqrt{y}+1/\sqrt{y})\Big)dy$$
$$=2^{\frac{1}{2}}\rho^{-1}\sqrt{\pi}\int_0^\infty y^{\nu-1}\exp\Big(-\tfrac{1}{2}\rho(\sqrt{y}+1/\sqrt{y})\Big)dy.$$

We replace y by y^2, and again use (1.1.17), getting (2.6.7).

We put $\nu=ir$, $r\in\mathbb{R}$, in (2.6.7), and appeal to (2.6.1). We have

$$\int_{-\infty}^\infty r\sinh(\pi r)K_{2ir}(\rho)K_{ir}(x)dr=2^{-\frac{5}{2}}\pi^{\frac{3}{2}}\rho x^{-\frac{1}{2}}\exp(-x-\rho^2/(8x)). \quad (2.6.8)$$

We multiply both sides by $e^{-x}x^{s-1}$ and integrate over the positive real axis with respect to x. Using (1.1.17) we get

$$\int_0^\infty e^{-x}x^{s-1}\int_{-\infty}^\infty r\sinh(\pi r)K_{2ir}(\rho)K_{ir}(x)drdx = 2^{-2s-\frac{1}{2}}\pi^{\frac{3}{2}}\rho^{s+\frac{1}{2}}K_{s-\frac{1}{2}}(\rho).$$

When $\mathrm{Re}\, s > 0$, the left side is absolutely convergent because of (1.3.45)–(1.3.47). Hence, exchanging the order of integration, we have

$$\int_{-\infty}^\infty r\sinh(\pi r)K_{2ir}(\rho)\Gamma(s+ir)\Gamma(s-ir)dr$$
$$= \pi\Gamma(s+\tfrac{1}{2})(\rho/2)^{s+\frac{1}{2}}K_{s-\frac{1}{2}}(\rho) \quad (\mathrm{Re}\, s > 0), \tag{2.6.9}$$

where (2.1.21) has been used.

We now suppose

$$\mathrm{Re}\,\omega_1,\ \mathrm{Re}\,\omega_2,\ \mathrm{Re}\,\omega_3 > 0. \tag{2.6.10}$$

We put $s = \omega_3$ in (2.6.9), multiply both sides of the identity by the factor $K_{\omega_1-\omega_2}(\rho)(\rho/2)^{\omega_1+\omega_2-1}$, and integrate over the positive real axis with respect to ρ. We then invoke the fact that if $\mathrm{Re}\, s > |\mathrm{Re}\,\mu| + |\mathrm{Re}\,\nu|$

$$\int_0^\infty K_\mu(x)K_\nu(x)(x/2)^{s-1}dx = \frac{1}{4}\Gamma(\tfrac{1}{2}(s+\mu+\nu))\Gamma(\tfrac{1}{2}(s+\mu-\nu))$$
$$\times\Gamma(\tfrac{1}{2}(s-\mu+\nu))\Gamma(\tfrac{1}{2}(s-\mu-\nu))/\Gamma(s), \tag{2.6.11}$$

the proof of which is to be given later. Thus (2.6.9) implies that

$$\int_0^\infty K_{\omega_1-\omega_2}(\rho)(\rho/2)^{\omega_1+\omega_2-1}$$
$$\times\int_{-\infty}^\infty r\sinh(\pi r)K_{2ir}(\rho)\Gamma(\omega_3+ir)\Gamma(\omega_3-ir)drd\rho$$
$$= \frac{\pi}{4}\Gamma(\omega_1+\tfrac{1}{2})\Gamma(\omega_2+\tfrac{1}{2})\Gamma(\omega_3+\tfrac{1}{2})\Gamma(\omega_1+\omega_3)$$
$$\times\Gamma(\omega_2+\omega_3)/\Gamma(\omega_1+\omega_2+\omega_3+\tfrac{1}{2}),$$

provided (2.6.10) holds. The left side converges absolutely. We exchange the order of integration, and compute the resulting inner integral with (2.6.11). Then we are led to (2.6.5) on the present assumption. By analytic continuation we end the proof of (2.6.5).

Next we consider (2.6.6). In (2.6.11) we put $\mu = 2ir$, $\nu = \omega_1 - \omega_2$, and $s = \omega_1 + \omega_2$; we again assume (2.6.10). Then by (2.6.1) with r being replaced by $2r$ we get

$$\frac{1}{2\pi^2}\int_{-\infty}^\infty r\sinh(2\pi r)K_{2ir}(x)\Gamma(\omega_1+ir)\Gamma(\omega_2+ir)\Gamma(\omega_1-ir)\Gamma(\omega_2-ir)dr$$
$$= \Gamma(\omega_1+\omega_2)K_{\omega_1-\omega_2}(x)(x/2)^{\omega_1+\omega_2}.$$

As a matter of fact, in order to use (2.6.1), we should assume that $\mathrm{Re}\,\omega_1$ and $\mathrm{Re}\,\omega_2$ are sufficiently large. But this condition can be eliminated afterward by analytic continuation. We multiply both sides of this identity by $(x/2)^{2\omega_3-1}$ and integrate over the positive real axis with respect to x. To compute the resulting right side we use the inversion of (2.5.2):

$$\int_0^\infty K_{2\nu}(x)(x/2)^{2s-1}dx = \frac{1}{2}\Gamma(s+\nu)\Gamma(s-\nu) \quad (\mathrm{Re}\,s > |\mathrm{Re}\,\nu|). \tag{2.6.12}$$

Thus we have

$$\frac{1}{\pi^2}\int_0^\infty (x/2)^{2\omega_3-1}\int_{-\infty}^\infty r\sinh(2\pi r)K_{2ir}(x)\Gamma(\omega_1+ir)\Gamma(\omega_2+ir) \\ \times\Gamma(\omega_1-ir)\Gamma(\omega_2-ir)drdx = \Gamma(\omega_1+\omega_2)\Gamma(\omega_2+\omega_3)\Gamma(\omega_3+\omega_1).$$

The double integral converges absolutely because of (1.3.45)–(1.3.47). We exchange the order of integration and evaluate the inner integral with (2.6.12). After analytic continuation of the result we end the proof of (2.6.6).

Hence what remains is to prove (2.6.11): By (2.5.2) we see that the left side is equal to

$$\frac{1}{2\pi i}\int_0^\infty K_\mu(y)\int_{(\beta)} 2^{2\xi-s}\Gamma(\xi+\tfrac{1}{2}\nu)\Gamma(\xi-\tfrac{1}{2}\nu)y^{s-2\xi-1}d\xi dy.$$

This double integral converges absolutely provided $\mathrm{Re}\,s > 2\beta + |\mathrm{Re}\,\mu| > |\mathrm{Re}\,\mu| + |\mathrm{Re}\,\nu|$. Exchanging the order of integration and computing the inner integral with (2.6.12) we get the expression

$$\frac{1}{8\pi i}\int_{(\beta)}\Gamma(\xi+\tfrac{1}{2}\nu)\Gamma(\xi-\tfrac{1}{2}\nu)\Gamma(\tfrac{1}{2}(s+\mu)-\xi)\Gamma(\tfrac{1}{2}(s-\mu)-\xi)d\xi.$$

Then we invoke Barnes' formula:

$$\frac{1}{2\pi i}\int_{-i\infty}^{i\infty}\Gamma(\xi+a)\Gamma(\xi+b)\Gamma(c-\xi)\Gamma(d-\xi)d\xi \\ = \frac{\Gamma(a+c)\Gamma(a+d)\Gamma(b+c)\Gamma(b+d)}{\Gamma(a+b+c+d)} \tag{2.6.13}$$

provided a, b, c, d are such that the path separates the poles of the first two Γ-factors and those of the other two Γ-factors in the integrand to the left and to the right, respectively. This obviously yields the formula (2.6.11), and we end the proof of Lemma 2.7.

Next we employ (2.6.5) to transform (2.1.2) into

Lemma 2.8 *If* $\operatorname{Re} s_1, \operatorname{Re} s_2 > \frac{3}{4}$ *then we have*

$$\begin{aligned}\langle P_m(\cdot, s_1), P_n(\cdot, \overline{s_2})\rangle &= \delta_{m,n}\Gamma(s_1+s_2-1)(4\pi m)^{1-s_1-s_2}\\ &+ \pi\frac{(4\pi)^{1-s_1-s_2}m^{\frac{1}{2}-s_1}n^{\frac{1}{2}-s_2}}{\Gamma(s_1)\Gamma(s_2)}\\ &\times\sum_{l=1}^{\infty}\frac{1}{l}S(m,n\,;l)\frac{2i}{\pi}\int_{-\infty}^{\infty} rJ_{2ir}\left(\frac{4\pi}{l}\sqrt{mn}\right)\Theta(s_1,s_2;r)dr, \end{aligned}\tag{2.6.14}$$

where Θ *is as in* (2.1.15).

Proof Comparing (2.1.2) with (2.6.14) we see that it is sufficient to show that for $\operatorname{Re} s_1, \operatorname{Re} s_2 > \frac{1}{2}$

$$W(x;s_1,s_2) = -\frac{(x/2)^{1-2s_1}}{\pi i\Gamma(s_1)\Gamma(s_2)\Gamma(s_1+s_2-1)}\int_{-\infty}^{\infty} rJ_{2ir}(x)\Theta(s_1,s_2;r)dr. \tag{2.6.15}$$

For this purpose we replace the three suffixed parameters in (2.6.5) as follows: $\omega_1 = s_1 - \frac{1}{2}$, $\omega_2 = s_2 - \frac{1}{2}$, and $\omega_3 = \eta - s_1 + \frac{1}{2}$. We assume temporarily that

$$\operatorname{Re} s_1, \operatorname{Re} s_2 > \tfrac{1}{2}\,;\ \operatorname{Re}(\eta - s_1) > -\tfrac{1}{2}. \tag{2.6.16}$$

Then we get, after a little rearrangement,

$$\frac{\Gamma(\eta)\Gamma(\eta+s_2-s_1)}{\Gamma(\eta+s_2)\Gamma(s_1-\eta)} = \frac{i\pi^{-1}U(s_1,s_2,\eta)}{\Gamma(s_1)\Gamma(s_2)\Gamma(s_1+s_2-1)}, \tag{2.6.17}$$

where

$$\begin{aligned}U(s_1,s_2,\eta) &= \frac{1}{\pi}\sin(\pi(s_1-\eta))\\ &\times\int_{(0)} \xi\sin(\pi\xi)\Theta(s_1,s_2;i\xi)\Gamma(\eta-s_1+\tfrac{1}{2}+\xi)\Gamma(\eta-s_1+\tfrac{1}{2}-\xi)d\xi.\end{aligned}$$

To remove the last condition in (2.6.16) which is in disaccord with those given in Lemma 2.1 we note that the functional equation $\Gamma(s)\Gamma(1-s) = \pi/\sin(\pi s)$ gives

$$\begin{aligned}&\sin(\pi(s_1-\eta))\Gamma(\eta-s_1+\tfrac{1}{2}+\xi)\Gamma(\eta-s_1+\tfrac{1}{2}-\xi)\\ &\qquad= -\frac{\pi}{2\sin(\pi\xi)}\left[\frac{\Gamma(\eta-s_1+\frac{1}{2}+\xi)}{\Gamma(s_1-\eta+\frac{1}{2}+\xi)} - \frac{\Gamma(\eta-s_1+\frac{1}{2}-\xi)}{\Gamma(s_1-\eta+\frac{1}{2}-\xi)}\right],\end{aligned}$$

which implies

$$U(s_1,s_2,\eta) = -\int_{(0)} \xi\Theta(s_1,s_2;i\xi)\frac{\Gamma(\eta-s_1+\frac{1}{2}+\xi)}{\Gamma(s_1-\eta+\frac{1}{2}+\xi)}d\xi$$

(cf. (2.4.17)). Providing

$$\mathrm{Re}\, s_1,\ \mathrm{Re}\, s_2 > \tfrac{1}{2} + \beta \quad (\beta > 0), \tag{2.6.18}$$

we have

$$U(s_1, s_2, \eta) = \int_{(\beta)} \cdots \, d\xi. \tag{2.6.19}$$

Hence we see that the relation (2.6.17) is valid if (2.6.18) as well as

$$\mathrm{Re}\,(\eta - s_1) > -\tfrac{1}{2} - \beta \tag{2.6.20}$$

hold. This is a relaxation of (2.6.16).

Now we have, by (2.6.17),

$$W(s_1, s_2; x) = \frac{1}{2\pi^2 \Gamma(s_1)\Gamma(s_2)\Gamma(s_1 + s_2 - 1)} \times \int_{(\alpha)} U(s_1, s_2, \eta)(x/2)^{-2\eta} d\eta \quad (\alpha > 0),$$

where (2.6.18)–(2.6.20) are incorporated. On the right side is actually a double integral. It is, by Stirling's formula,

$$\ll \int_{(\alpha)} \int_{(\beta)} (1 + |\xi|)^{2\mathrm{Re}\,(s_1+s_2)-3} (1 + |\eta + \xi|)^{\alpha+\beta-\mathrm{Re}\, s_1} \times (1 + |\eta - \xi|)^{\alpha-\beta-\mathrm{Re}\, s_1} \exp(-2\pi|\xi| - \tfrac{1}{2}\pi(|\eta + \xi| - |\eta - \xi|))|d\xi d\eta| < +\infty$$

provided $\mathrm{Re}\, s_1 > \frac{1}{2} + \alpha$. Thus, on the condition that

$$\tfrac{1}{2} + \alpha + \beta > \mathrm{Re}\, s_1 > \tfrac{1}{2} + \alpha, \quad \mathrm{Re}\, s_2 > \tfrac{1}{2} + \beta$$

with arbitrary $\alpha > \beta > 0$, we have

$$\int_{(\alpha)} U(s_1, s_2, \eta)(x/2)^{-2\eta} d\eta = -\int_{(\beta)} \xi \Theta(s_1, s_2; i\xi) \int_{(\alpha)} \frac{\Gamma(\eta - s_1 + \frac{1}{2} + \xi)}{\Gamma(s_1 - \eta + \frac{1}{2} + \xi)} (x/2)^{-2\eta} d\eta d\xi.$$

By (2.2.17) the inner integral is equal to

$$2\pi i J_{2\xi}(x)(x/2)^{1-2s_1}.$$

This and a little additional consideration prove (2.6.15), and we end the proof of the lemma.

We then turn to the 'opposite-sign' case:

Lemma 2.9 *Under the same supposition as in the previous lemma we have*

$$\langle P_m(\cdot, s_1), \overline{P_n(\cdot, s_2)}\rangle = \pi \frac{(4\pi)^{1-s_1-s_2} m^{\frac{1}{2}-s_1} n^{\frac{1}{2}-s_2}}{\Gamma(s_1)\Gamma(s_2)}$$
$$\times \sum_{l=1}^{\infty} \frac{1}{l} S(m, -n\,; l) \frac{2}{\pi^2} \int_{-\infty}^{\infty} r \sinh(2\pi r) K_{2ir}\left(\frac{4\pi}{l}\sqrt{mn}\right) \Theta(s_1, s_2; r) dr. \tag{2.6.21}$$

Proof This time we transform (2.5.1). The identity (2.6.6) gives

$$\Gamma(s_1 + s_2 - 1)\Gamma(\eta + s_1)\Gamma(\eta + s_2)$$
$$= \frac{i}{2\pi^2} \int_{(0)} \xi \sin(2\pi\xi)\Theta(s_1, s_2; i\xi)\Gamma(\eta + \tfrac{1}{2} + \xi)\Gamma(\eta + \tfrac{1}{2} - \xi) d\xi.$$

Hence (2.5.2) yields

$$\frac{1}{2\pi i}\Gamma(s_1 + s_2 - 1) \int_{(0)} \Gamma(\eta + s_1)\Gamma(\eta + s_2)(x/2)^{-2\eta} d\eta$$
$$= \frac{i}{2\pi^2} \int_{(0)} \xi \sin(2\pi\xi)\Theta(s_1, s_2; -i\xi)$$
$$\times \frac{1}{2\pi i} \int_{(0)} \Gamma(\eta + \tfrac{1}{2} + \xi)\Gamma(\eta + \tfrac{1}{2} - \xi)(x/2)^{-2\eta} d\eta d\xi$$
$$= \frac{x}{2\pi^2} \int_{-\infty}^{\infty} r \sinh(2\pi r)\Theta(s_1, s_2; r) K_{2ir}(x) dr,$$

which obviously ends the proof.

Here it should be observed that the assertions (2.6.14) and (2.6.21) are exactly the same as the results of applying the trace formulas (2.3.16) and (2.5.6) to (2.1.14) and (2.5.3), respectively, although the application should be made on the assumption $\operatorname{Re} s_1$, $\operatorname{Re} s_2 > 1$, and analytic continuation is required. What is novel in the above procedure is that the relations (2.6.14) and (2.6.21) have been obtained without the trace formulas, and this has been made possible by (2.6.5)–(2.6.6).

Now a new proof of the trace formulas (2.3.16) and (2.5.6) follows: From (2.1.14), (2.5.3), (2.6.14), and (2.6.21) we have, for $mn \neq 0$,

$\operatorname{Re} s_1, \operatorname{Re} s_2 > \frac{3}{4}$,

$$\begin{aligned}\sum_{j=1}^{\infty} &\overline{\rho_j(m)}\rho_j(n)\Theta(s_1, s_2\,; \kappa_j)\\ &+ \frac{1}{\pi}\int_{-\infty}^{\infty} \frac{\sigma_{2ir}(m)\sigma_{2ir}(n)\cosh(\pi r)\Theta(s_1, s_2\,; r)}{|mn|^{ir}|\zeta(1+2ir)|^2} dr\\ &= \delta_{m,n}\pi^{-1}\Gamma(s_1)\Gamma(s_2)\Gamma(s_1+s_2-1)\\ &+ \sum_{l=1}^{\infty} \frac{1}{l} S(m, n\,; l)\frac{2i}{\pi}\int_{-\infty}^{\infty} r\tilde{J}_{2ir}\left(\frac{4\pi}{l}\sqrt{|mn|}\right)\Theta(s_1, s_2; r)dr, \qquad (2.6.22)\end{aligned}$$

where $\tilde{J}_\nu$ is equal to J_ν if $mn > 0$, and to I_ν if $mn < 0$; and we have used (1.1.25). We specialize this by (2.3.3). We then multiply both sides of the resulting identity by the factor $\cosh(\pi t)f(t)$ with f satisfying (2.3.15), and integrate with respect to t over the real axis. The rest of the argument is the same as the proof of Theorem 2.2, save for the use of the expression

$$I_\nu(x) = \frac{(x/2)^\nu}{\sqrt{\pi}\Gamma(\nu+\frac{1}{2})}\int_{-1}^{1}(1-u^2)^{\nu-\frac{1}{2}}\cosh(xu)du \quad (x > 0, \operatorname{Re}\nu > -\tfrac{1}{2}) \qquad (2.6.23)$$

when treating the opposite-sign case.

Still another proof is available. Perhaps this argument is more interesting than the last one for it yields a genuine improvement upon the results (2.3.16) and (2.5.6). That is, we are able to replace the condition (2.3.15) by

$$\left.\begin{aligned}&f(r) \text{ is regular in the strip } |\operatorname{Im} r| \le \tfrac{1}{4}+\delta, \text{ and there}\\ &\qquad f(r) = f(-r),\ |f(r)| \ll (1+|r|)^{-2-\delta},\end{aligned}\right\} \qquad (2.6.24)$$

where δ is an arbitrary small positive constant. Also the argument contains a new technical point as well. This is the use of the dual of the inversion formula (2.6.1). So we begin our third proof of the trace formulas with the assertion

Lemma 2.10 *Let $k(\xi)$ be an even function which is regular in the horizontal strip $|\operatorname{Im}\xi| \le \alpha$ for a certain $\alpha > 0$, and let us assume that $k(\xi) = O(e^{-2\pi|\xi|})$ there. Then we have*

$$k(t) = \pi^{-2}\int_0^{\infty} K_{it}(v)v^{-1}\int_{-\infty}^{\infty}\xi\sinh(\pi\xi)K_{i\xi}(v)k(\xi)d\xi dv \quad (|\operatorname{Im} t| < \alpha). \qquad (2.6.25)$$

Proof The decay condition on k could of course be replaced by a less stringent one. Let us denote the right side by $k^*(t)$. We shall show first that $k^*(t)$ is regular for $|\mathrm{Im}\, t| < \alpha$. To this end let $k_0(v)$ be the inner integral. It is easy to see that $k_0(v)$ is of exponential decay as v tends to positive infinity; we need only to take the absolute value of the integrand. On the other hand, when v is small, we use (1.1.25). We have

$$k_0(v) = \pi i \int_{-\infty}^{\infty} \xi I_{i\xi}(v) k(\xi) d\xi.$$

Shifting the path to $\mathrm{Im}\, \xi = -\alpha$, we get $k_0(v) = O(v^\alpha)$ in view of either (1.1.24) or (2.6.23). These estimates of $k_0(v)$ confirm the above claim. Next we shall show that $k^*(t) = k(t)$ for $t \in \mathbb{R}$, which will obviously finish the proof. So let t be real, and let $k^*(s,t)$ be the result of replacing the factor v^{-1} in the integrand in (2.6.25) by v^{s-1} with a complex s. We note that $k^*(s,t)$ is regular for $\mathrm{Re}\, s > -\alpha$, and that when $\mathrm{Re}\, s > 0$ the double integral defining $k^*(s,t)$ is absolutely convergent. Thus we have, by means of (2.6.11),

$$\begin{aligned} k^*(t) &= \lim_{s \to 0^+} k^*(s,t) \\ &= \frac{1}{8\pi^2} \lim_{s \to 0^+} \frac{2^s}{\Gamma(s)} \int_{-\infty}^{\infty} \xi \sinh(\pi\xi) \Gamma(\tfrac{1}{2}(s + i\xi + it)) \Gamma(\tfrac{1}{2}(s + i\xi - it)) \\ &\qquad \times \Gamma(\tfrac{1}{2}(s - i\xi + it)) \Gamma(\tfrac{1}{2}(s - i\xi - it)) k(\xi) d\xi. \end{aligned} \tag{2.6.26}$$

We shift the path upwards slightly. Computing the residues at the simple poles $\xi = \pm t + is$, we readily find that $k^*(t) = k(t)$. This ends the proof of the lemma.

Returning to the proof of the trace formulas, we observe first that we may consider instead the case where we have $e^{-ar^2} h(r)$, $a > 0$, as the weight with $h(r)$ satisfying (2.6.24). This is because by virtue of the bound (2.3.2) the trace formulas with this modified weight function should hold uniformly for $a \geq 0$. We remark that in order to confirm the uniform convergence, with respect to a, of the Kloosterman-sum parts of these trace formulas we need to shift the line of integration involved there in the way given in the expression below.

Hence we may assume that $h(r)$, in the new context, is very rapidly decaying as $|r|$ tends to infinity in the strip $|\mathrm{Im}\, r| \leq \frac{1}{4} + \delta$. We then put $s_2 = 1$, $s_1 = s$ in (2.6.22). We have, for $mn \neq 0$, $\mathrm{Re}\, s > \frac{3}{4} + \varepsilon$ with a small

$\varepsilon > 0$,

$$\sum_{j=1}^{\infty} \frac{\overline{\rho_j(m)}\rho_j(n)}{\cosh \pi\kappa_j}\Gamma(s-\tfrac{1}{2}+i\kappa_j)\Gamma(s-\tfrac{1}{2}-i\kappa_j)$$
$$+\frac{1}{\pi}\int_{-\infty}^{\infty}\frac{\sigma_{ir}(m)\sigma_{ir}(n)}{|mn|^{ir}|\zeta(1+2ir)|^2}\Gamma(s-\tfrac{1}{2}+ir)\Gamma(s-\tfrac{1}{2}-ir)dr$$
$$=\pi^{-1}\delta_{m,n}\Gamma^2(s)+\sum_{l=1}^{\infty}\frac{1}{l}S(m,n;l)$$
$$\times\frac{2i}{\pi}\int_{[\varepsilon]}\frac{r}{\cosh \pi r}\tilde{J}_{2ir}\left(\frac{4\pi}{l}\sqrt{|mn|}\right)\Gamma(s-\tfrac{1}{2}+ir)\Gamma(s-\tfrac{1}{2}-ir)dr.$$

Here $[\varepsilon]$ indicates that the path is $\operatorname{Im} r = -\frac{1}{4}-\varepsilon$. The shift of path involved here is to gain absolute convergence in this and subsequent expressions. We multiply both sides by the factor $(v/2)^{1-2s}$, $v > 0$, and integrate with respect to s over the line $\operatorname{Re} s = 1$, say. It is easy to check the absolute convergence, and by (2.5.2) we have, for any $v > 0$,

$$\sum_{j=1}^{\infty}\frac{\overline{\rho_j(m)}\rho_j(n)}{\cosh \pi\kappa_j}K_{2i\kappa_j}(v)$$
$$+\frac{1}{\pi}\int_{-\infty}^{\infty}\frac{\sigma_{ir}(m)\sigma_{ir}(n)}{|mn|^{ir}|\zeta(1+ir)|^2}K_{2ir}(v)dr=\frac{1}{2\pi}\delta_{m,n}vK_0(v)$$
$$+\sum_{l=1}^{\infty}\frac{1}{l}S(m,n;l)\frac{2i}{\pi}\int_{[\varepsilon]}\frac{r}{\cosh \pi r}\tilde{J}_{2ir}\left(\frac{4\pi}{l}\sqrt{|mn|}\right)K_{2ir}(v)dr.$$

We are now at a position to appeal to (2.6.25). We multiply both sides of the above by the factor

$$g(v)=\frac{1}{\pi^2 v}\int_{-\infty}^{\infty}\xi\sinh(\pi\xi)K_{i\xi}(v)h(\xi/2)d\xi,$$

and integrate with respect to v over the positive real axis. Supposing the necessary absolute convergence, we get, by (2.6.25), both (2.3.16) and (2.5.6) for the present choice of h, save for the verification of

$$\int_0^{\infty}vK_0(v)g(v)dv=2\pi^{-1}\int_{-\infty}^{\infty}r\tanh(\pi r)h(r)dr.$$

But this is a consequence of (2.6.11). Thus it remains to examine the absolute convergence. We have $g(v) \ll v^{-\frac{1}{2}+2\delta}$ as $v \to 0^+$ and $\ll e^{-v}$ as $v \to +\infty$. Also we have $K_{2ir}(v) \ll v^{2\operatorname{Im} r}e^{-|r|}$ uniformly in the relevant r, $|r| \geq 1$, and arbitrary $v \geq 0$, which follows from (1.1.24)–(1.1.25) and (1.3.45)–(1.3.47). These estimates of $g(v)$ and $K_{2ir}(v)$ readily imply absolute convergence on the spectral side. As to the part involving Kloosterman sums we have $\operatorname{Im} r = -\frac{1}{4} - \varepsilon$, and thus $\tilde{J}_{2ir}(4\pi l^{-1}|mn|^{\frac{1}{2}}) \ll e^{\pi|r|}l^{-\frac{1}{2}-2\varepsilon}$ uniformly in r and l. Hence the double integral over v and r converges absolutely and is $O(l^{-\frac{1}{2}-2\varepsilon})$, provided $\varepsilon = \frac{1}{2}\delta$, say. This obviously ends our third proof of the trace formulas. We could of course combine (2.6.22) and (2.6.26) to attain a shorter argument, however.

2.7 A flight

So far we have been moving around in the upper half plane, looking at the real axis as the horizon. The aim of this section is to indicate a way of flying over the plane $\mathbb{C}$, regarding it as the celestial sphere: We shall show an extension of the argument of the previous sections to the three dimensional non-Euclidean space $\mathcal{H}^3$. The main result is an analogue of the trace formula (2.3.16) for the Gaussian number field. This should have various applications related to the topics treated in the later chapters. Our immediate motivation is, however, to provide readers with a wider perspective of the spectral method. Thus we shall limit ourselves to an expository account. For the sake of notational simplicity we shall employ old symbols to designate corresponding new concepts; hence it should be understood that the notation is effective only within this section.

We need first to explain briefly the basics of the structure of the space $\mathcal{H}^3$: The points of $\mathcal{H}^3$ are denoted by $z = (x, y)$ with $x = x_1 + x_2 i$ $(x_1, x_2 \in \mathbb{R})$ and $y > 0$. Algebraically $\mathcal{H}^3$ is embedded in the Hamiltonian algebra so that we have $\mathcal{H}^3 \ni z = x + y\jmath$ with $\jmath^2 = -1$ and $i\jmath = -\jmath i$. These two notations for the points will be used interchangeably. The metric is $((dx_1)^2 + (dx_2)^2 + (dy)^2)^{\frac{1}{2}}/y$, the corresponding volume element is $d\mu(z) = y^{-3}dx_1dx_2dy$, and the non-Euclidean Laplacian is

$$\Delta = -y^2\big((\partial/\partial x_1)^2 + (\partial/\partial x_2)^2 + (\partial/\partial y)^2\big) + y(\partial/\partial y). \tag{2.7.1}$$

As to the motions of $\mathcal{H}^3$, let $\mathbb{T}(\mathcal{H}^3)$ be the set of maps

$$\left\{z \mapsto (az+b)\cdot(lz+h)^{-1} \,:\, a,b,l,h \in \mathbb{C} \text{ with } ah - bl = 1\right\}, \tag{2.7.2}$$

where the algebraic operation is that of the Hamiltonian. In the coordinate notation this generic map is

$$z = (x, y) \mapsto \left(\frac{(ax+b)\overline{(lx+h)} + a\bar{l}y^2}{|lx+h|^2 + |ly|^2}, \frac{y}{|lx+h|^2+|ly|^2}\right). \tag{2.7.3}$$

With respect to the motions in $\mathbb{T}(\mathcal{H}^3)$, the metric, the volume element, and the Laplacian are all invariant. The concept corresponding to the full modular group is the Picard group Γ composed of those elements in (2.7.2) with $a, b, l, h \in \mathbb{Z}[i]$, i.e., Gaussian integers. It is known that Γ acts discontinuously over $\mathcal{H}^3$, having the fundamental domain

$$\mathcal{F} = \left\{z : x_1 \le \tfrac{1}{2},\ x_2 \le \tfrac{1}{2},\ x_1 + x_2 \ge 0,\ x_1^2 + x_2^2 + y^2 \ge 1\right\}. \tag{2.7.4}$$

We may take $\mathcal{F}$ for a three dimensional manifold as before. Also it should be noted that the volume of $\mathcal{F}$ is finite.

Now the set of all Γ-invariant functions over $\mathcal{H}^3$ which are square integrable with respect to $d\mu$ over $\mathcal{F}$ constitutes the Hilbert space $L^2(\mathcal{F}, d\mu)$ equipped with the inner-product

$$\langle f_1, f_2\rangle = \int_{\mathcal{F}} f_1(z)\overline{f_2(z)}d\mu(z).$$

The spectral resolution of Δ over $L^2(\mathcal{F}, d\mu)$ is analogous to Theorem 1.1. The non-trivial discrete spectrum of Δ is denoted by $\left\{\lambda_j = 1 + \kappa_j^2 : j = 1, 2, \ldots\right\}$ in non-decreasing order, and the corresponding orthonormal system of eigenfunctions by $\left\{\psi_j\right\}$. It is known that the number of λ_j's is infinite and that $\kappa_j > 0$. We have the Fourier expansion

$$\psi_j(z) = y \sum_{\substack{n\in\mathbb{Z}[i]\\ n\neq 0}} \rho_j(n) K_{i\kappa_j}(2\pi|n|y)e([n, x]).$$

Here K_ν is the K-Bessel function of order ν, $e(a) = \exp(2\pi i a)$ as before, and $[n, x] = \mathrm{Re}\,(n\overline{x})$. We also need to introduce the Eisenstein series

$$E(z, s) = \sum_{\gamma\in\Gamma_\infty\backslash\Gamma} y^s(\gamma(z)),$$

where $y(z)$ denotes the third coordinate of z, and Γ_∞ is the stabilizer subgroup in Γ of the point at infinity. We note that

$$\Gamma_\infty = \Gamma_t \cup \imath\Gamma_t$$

where Γ_t is the translation group $\{z \mapsto z + b : b \in \mathbb{Z}[i]\}$, and $\imath$ is the involution $\imath(z) = -x + y\jmath$, which is the same as $z \mapsto (iz)\cdot(-i)^{-1}$ in the Hamiltonian sense. That is, we have

$$E(z, s) = \frac{1}{4}\sum_{\substack{l,h\in\mathbb{Z}[i]\\ (l,h)=1}} \frac{y^s}{(|lx + h|^2 + |ly|^2)^s}.$$

It is easy to check that the series converges absolutely for $\mathrm{Re}\, s > 2$. We have the Fourier expansion

$$E(z,s) = y^s + \frac{\pi}{s-1}\frac{\zeta_K(s-1)}{\zeta_K(s)} y^{2-s} + \frac{2\pi^s y}{\Gamma(s)\zeta_K(s)} \sum_{\substack{n\in\mathbb{Z}[i] \\ n\neq 0}} |n|^{s-1}\sigma_{1-s}(n) K_{s-1}(2\pi|n|y) e([n,x]),$$

where ζ_K is the Dedekind zeta-function of the Gaussian number field, and $\sigma_v(n) = \frac{1}{4}\sum_{d|n} |d|^{2v}$ ($d \in \mathbb{Z}[i]$). Thus $E(z,s)$ is meromorphic over $\mathbb{C}$ as a function of s, being regular for $\mathrm{Re}\, s \geq 1$ except for the simple pole at $s = 2$.

With this notation we may state the Parseval formula: For any $f_1, f_2 \in L^2(\mathcal{F}, d\mu)$ we have

$$\langle f_1, f_2\rangle = \sum_{j=0}^{\infty} \langle f_1, \psi_j\rangle \overline{\langle f_2, \psi_j\rangle} + \frac{1}{2\pi}\int_{-\infty}^{\infty} \mathcal{E}(t, f_1)\overline{\mathcal{E}(t, f_2)} dt. \tag{2.7.5}$$

Here $\psi_0 \equiv \pi(2\zeta_K(2))^{-\frac{1}{2}}$ and

$$\mathcal{E}(t,f) = \int_{\mathcal{F}} f(z) E(z, 1-it)\, d\mu(z),$$

where the integral is to be taken in the sense of the limit in mean similar to (1.1.46).

The proof of (2.7.5) can be carried out by following closely the argument developed in Chapter 1. To some extent it is simpler, for the free-space resolvent kernel of $\Delta + \alpha(\alpha - 2)$ has the explicit expression

$$\left(\tfrac{1}{2}(\sqrt{\varrho} + \sqrt{\varrho + 1})\right)^{2(1-\alpha)} (\varrho(\rho + 1))^{-\frac{1}{2}} \tag{2.7.6}$$

with

$$\varrho = \left[(x_1 - u_1)^2 + (x_2 - u_2)^2 + (y - v)^2\right]/(4yv) \qquad (z = (x, y),\ w = (u, v)).$$

Having said these we may now introduce our new trace formula:

Theorem 2.6 *Let*

$$S(m,n;l) = \sum_{\substack{v \bmod l \\ (v,l)=1}} e([m, v/l])\, e([n, v^*/l]) \quad (l, m, n \in \mathbb{Z}[i];\ vv^* \equiv 1 \bmod l)$$

be the Kloosterman sum for the Gaussian number field. Let us assume that the function $h(r)$, $r \in \mathbb{C}$, *is regular in the horizontal strip* $|\mathrm{Im}\, r| \le \frac{1}{2} + \delta$ *and satisfies*

$$h(r) = h(-r), \quad h(r) \ll (1 + |r|)^{-3-\delta}$$

for an arbitrary fixed $\delta > 0$. *Then we have, for any non-zero* $m, n \in \mathbb{Z}[i]$,

$$\sum_{j=1}^{\infty} \frac{\overline{\rho_j(m)}\rho_j(n)}{\sinh(\pi\kappa_j)} \kappa_j h(\kappa_j) + 2\pi \int_{-\infty}^{\infty} \frac{\sigma_{ir}(m)\sigma_{ir}(n)}{|mn|^{ir}|\zeta_K(1+ir)|^2} h(r)dr$$
$$= (\delta_{m,n} + \delta_{m,-n})\pi^{-2} \int_{-\infty}^{\infty} r^2 h(r)dr + \sum_{l \in \mathbb{Z}[i],\, l \neq 0} |l|^{-2} S(m,n;l)\check{h}(2\pi\varpi) \quad (2.7.7)$$

with $\varpi^2 = \overline{mn}/l^2$. *Here* $\delta_{m,n}$ *is the Kronecker delta, and*

$$\check{h}(t) = i \int_{-\infty}^{\infty} \frac{r^2}{\sinh(\pi r)} J_{ir}(t) J_{ir}(\bar{t}) h(r)dr.$$

Sketch of the proof First we should remark that the choice of the sign in ϖ is immaterial as can be seen from the power series expansion of the J-Bessel function.

We shall again work with the Poincaré series: For $m \in \mathbb{Z}[i]$ we put

$$P_m(z,s) = \sum_{\gamma \in \Gamma_t \backslash \Gamma} y^s(\gamma(z)) \exp\left\{-2\pi|m|y(\gamma(z)) + 2\pi i[m, x(\gamma(z))]\right\}$$
$$(\mathrm{Re}\, s > 2)$$

with an obvious convention. Expanding this into a double Fourier series with respect to the variables x_1, x_2, we get

$$P_m(z,s) = 2y^s \exp\left(-2\pi|m|y\right) \cos\left(2\pi[m,x]\right)$$
$$+ \frac{1}{2} y^{2-s} \sum_{n \in \mathbb{Z}[i]} e([n,x]) \sum_{\substack{l \in \mathbb{Z}[i] \\ l \neq 0}} |l|^{-2s} S(m,n;l) A_s(m,n;l;y), \quad (2.7.8)$$

where

$$A_s(m,n;l;y)$$
$$= \int_{-\infty}^{\infty}\int_{-\infty}^{\infty} (|\mu|^2+1)^{-s}$$
$$\times \exp\left(-2\pi i y[n,\mu] - \frac{2\pi|m|}{|l|^2(|\mu|^2+1)y} - \frac{2\pi i[l^{-2}, m\mu]}{(|\mu|^2+1)y}\right) d\mu_1 d\mu_2 \quad (2.7.9)$$

with $\mu = \mu_1 + i\mu_2$. Changing the variables by putting $\mu = ue^{i\theta}$, we also have

$$A_s(m,n;l;y)$$
$$= 2\pi \int_0^\infty \frac{u}{(u^2+1)^s} J_0\left(2\pi u\left|ny + \frac{\overline{m}}{l^2(u^2+1)y}\right|\right) \exp\left(-\frac{2\pi|m|}{|l|^2(u^2+1)y}\right) du.$$

In (2.7.8), which obviously corresponds to (1.1.6), we have

$$S(m,n;l) \ll |l||(m,n,l)|\sigma_0(l); \tag{2.7.10}$$

and, uniformly in s, l, and y,

$$A_s(m,n;l;y) \ll \exp(-|n|y),$$

as can be seen by shifting appropriately the two paths in (2.7.9). Hence $P_m(z,s)$ is an element of $L^2(\mathfrak{F}, d\mu)$ whenever $m \neq 0$ and $\mathrm{Re}\, s > \frac{3}{2}$.

Moving to the situation analogous to the earlier part of this chapter, we consider the inner-product $\langle P_m(\cdot, s_1), P_n(\cdot, \overline{s_2})\rangle$ in the space $L^2(\mathfrak{F}, d\mu)$, where $mn \neq 0$; and $\mathrm{Re}\, s_1, \mathrm{Re}\, s_2 > \frac{3}{2}$. By the unfolding method we have, for $\mathrm{Re}\, s_1$, $\mathrm{Re}\, s_2 > 2$,

$$\langle P_m(\cdot, s_1), P_n(\cdot, \overline{s_2})\rangle = (\delta_{m,n} + \delta_{m,-n})\Gamma(s_1+s_2-2)(4\pi|m|)^{2-s_1-s_2}$$
$$+ \pi(|m|/|n|)^{\frac{1}{2}(s_2-s_1)} \sum_{\substack{l \in \mathbb{Z}[i] \\ l \neq 0}} |l|^{-s_1-s_2} S(m,n;l) B(2\pi\sqrt{|mn|}/|l|, \vartheta_0; s_1, s_2),$$

where $\vartheta_0 = \arg \varpi$ with ϖ as above, and

$$B(p, \vartheta; s_1, s_2) = \int_0^\infty y^{s_2-s_1-1} C(p, \vartheta; \tfrac{1}{2}(s_1+s_2); y) dy$$

with $|\vartheta| \le \frac{1}{2}\pi$ and

$$C(p, \vartheta; \tau; y) = \int_0^\infty \frac{u}{(u^2+1)^\tau} \exp\left(-\frac{p(y+y^{-1})}{\sqrt{u^2+1}}\right)$$
$$\times J_0\left(\frac{pu}{\sqrt{u^2+1}}\left|ye^{i\vartheta} + (ye^{i\vartheta})^{-1}\right|\right) du.$$

To separate the variables in this integral we use the Mellin transform

$$\int_0^\infty a^{\eta-1} e^{-a} J_0(ab) da = \Gamma(\eta) F(\tfrac{1}{2}\eta, \tfrac{1}{2}(\eta+1); 1; -b^2). \tag{2.7.11}$$

Inverting this relation, and invoking the Mellin–Barnes formula for the hypergeometric function F, we are led to the expression

$$\exp(\cdots)J_0(\cdots) = -\frac{1}{2\pi^2}\int_{(\alpha)}\left(\frac{p(y+y^{-1})}{\sqrt{u^2+1}}\right)^{-2\eta} \times \int_{(\beta)}\frac{\Gamma(2\xi+2\eta)\Gamma(-\xi)}{\Gamma(\xi+1)}\left(1-\left(\frac{2\sin\vartheta}{y+y^{-1}}\right)^2\right)^{\xi}(u/2)^{2\xi}\,d\xi d\eta,$$

where $\alpha > 0$, $\beta < 0$ are small while satisfying $\alpha+\beta > 0$. This double integral is, however, not absolutely convergent. To gain the absolute convergence we shift the contour $\operatorname{Re}\xi = \beta$ to $\operatorname{Re}\xi = \beta'$ with a small $\beta' > 0$. The pole at $\xi = 0$ contributes a term which does not cause any trouble. We insert the resulting expression into the integral for $C(p,\vartheta;\tau;y)$. The triple integral thus obtained is absolutely convergent provided $\operatorname{Re}\tau > 1$. We perform the u-integral first, getting

$$C(p,\vartheta;\tau;y) = R - \frac{1}{4\pi^2}\int_{(\alpha)}\int_{(\beta')}\frac{\Gamma(2\xi+2\eta)\Gamma(\tau-\xi-\eta-1)\Gamma(-\xi)}{\Gamma(\tau-\eta)} \times (p(y+y^{-1}))^{-2\eta}\left(1-\left(\frac{2\sin\vartheta}{y+y^{-1}}\right)^2\right)^{\xi}2^{-2\xi}\,d\xi d\eta,$$

where R is the contribution of the pole at $\xi = 0$. We insert this into the integral for $B(p,\vartheta;s_1,s_2)$. The triple integral arising is absolutely convergent, provided $|\operatorname{Re}(s_1-s_2)| < 2\alpha$ which we shall assume for a while. We arrange the order of integration by making the ξ-integral inner, the y-integral middle, and the η-integral outer. To compute the inner integral we expand the factor $(1-(2\sin\vartheta/(y+y^{-1}))^2)^{\xi}$ into a binomial series. The ξ-integral can be performed inside this expansion; the termwise integration can be accomplished with (2.6.13). The resulting series converges absolutely, and the y-integral can be performed termwise. Then we find that

$$B(p,\vartheta;s_1,s_2) = -i\frac{\Gamma(s_1+s_2-2)}{2^{s_1+s_2-1}\sqrt{\pi}}\int_{(\alpha)}\frac{p^{-2\eta}}{\Gamma(\frac{1}{2}(s_1+s_2)-\eta)} \times \sum_{\nu=0}^{\infty}\frac{\Gamma(\eta+\frac{1}{2}(s_1-s_2)+\nu)\Gamma(\eta+\frac{1}{2}(s_2-s_1)+\nu)}{\Gamma(\nu+1)\Gamma(\eta+\frac{1}{2}(s_1+s_2-1)+\nu)}(\sin\vartheta)^{2\nu}\,d\eta. \tag{2.7.12}$$

This integral converges absolutely but the whole expression does not. To attain the same effect as the exchange of the order of the sum and the integral we apply, to the last sum, the integral representation (3.3.40) of the hypergeometric function. Then we have an absolutely convergent double integral. The η-integral can be taken as inner, and we get, after a change of variable,

$$B(p,\vartheta;s_1,s_2) = 8\sqrt{\pi}\frac{p^{1-s_2}\Gamma(s_1+s_2-2)}{2^{s_1+s_2}\Gamma(s_1-\frac{1}{2})}\int_0^\infty u^{2s_1-2}(u^2+1)^{\frac{1}{2}-s_2}$$
$$\times (u^2+(\cos\vartheta)^2)^{\frac{1}{2}(1-s_1)}J_{s_1-1}(2p\sqrt{u^2+(\cos\vartheta)^2})du.$$

Expressing the factor $(u^2+1)^{\frac{1}{2}-s_2}$ as an inverse Mellin integral, we are led to the situation where we may appeal to the integral formula:

$$\int_0^\infty \frac{J_\rho(\sqrt{u^2+a^2})}{(u^2+a^2)^{\frac{1}{2}\rho}}u^{2\mu+1}du = 2^\mu a^{\mu+1-\rho}\Gamma(\mu+1)J_{\rho-\mu-1}(a), \qquad (2.7.13)$$

where $a > 0$ and $\mathrm{Re}\left(\frac{1}{2}\rho - \frac{1}{4}\right) > \mathrm{Re}\,\mu > -1$. This procedure yields

$$B(p,\vartheta;s_1,s_2) = \frac{(2p)^{2-s_1-s_2}\Gamma(s_1+s_2-2)}{2i\sqrt{\pi}\Gamma(s_1-\frac{1}{2})\Gamma(s_2-\frac{1}{2})}$$
$$\times\int_{(\delta)}\Gamma(\xi)\Gamma(s_1-\tfrac{1}{2}-\xi)\Gamma(s_2-\tfrac{1}{2}-\xi)J_{\xi-\frac{1}{2}}(2p\cos\vartheta)\left(\frac{p}{\cos\vartheta}\right)^{\xi-\frac{1}{2}}d\xi,$$

where $0 < \delta < \min[\mathrm{Re}\,s_1, \mathrm{Re}\,s_2] - \frac{1}{2}$; and the above restriction on s_1, s_2 has been dropped. We replace the J-factor by the representation (2.2.8), and get, after a change of variable,

$$\int_{(\delta)}\cdots = \frac{2}{\sqrt{\pi}}\int_0^{\frac{1}{2}\pi}\cos(2p\cos\vartheta\cos\tau)$$
$$\times\int_{(\delta)}\Gamma(s_1-\tfrac{1}{2}-\xi)\Gamma(s_2-\tfrac{1}{2}-\xi)(p\sin\tau)^{2\xi-1}d\xi d\tau.$$

The inner integral is essentially a value of the K-Bessel function of order $s_1 - s_2$. But we rather appeal to (2.6.6). In it we put $\omega_1 = s_1 - 1$, $\omega_2 = s_2 - 1$, $\omega_3 = \frac{1}{2} - \xi$ assuming that $\mathrm{Re}\,\xi = \delta$ is small and that $\mathrm{Re}\,s_1$, $\mathrm{Re}\,s_2 > 1$. Then the last expression is transformed into

$$-\frac{4\pi^{-\frac{3}{2}}}{\Gamma(s_1+s_2-2)}\int_{(0)}\lambda\sin(2\pi\lambda)\Theta(s_1-\tfrac{1}{2},s_2-\tfrac{1}{2};\lambda)$$
$$\times\int_0^{\frac{1}{2}\pi}\cos(2p\cos\vartheta\cos\tau)K_{2\lambda}(2p\sin\tau)d\tau d\lambda, \qquad (2.7.14)$$

where Θ is defined by (2.1.15). The inner integral is expressible in terms of Bessel functions. This is a consequence of the formula

$$\int_0^{\frac{1}{2}\pi}\cos(a\cos\vartheta)J_{2\lambda}(b\sin\vartheta)d\vartheta$$
$$= \frac{1}{2}\pi J_\lambda\left(\tfrac{1}{2}(\sqrt{a^2+b^2}+a)\right)J_\lambda\left(\tfrac{1}{2}(\sqrt{a^2+b^2}-a)\right), \qquad (2.7.15)$$

where $a, b > 0$, $\mathrm{Re}\,\lambda > -\frac{1}{2}$. In fact, by analytic continuation, we may move b to the imaginary axis, and we find that the resulting integral formula and the definition (1.1.25) yield the following closed form for the inner integral of (2.7.14):

$$\int_0^{\frac{1}{2}\pi} \cos(\cdots)K_{2\lambda}(\cdots)d\tau$$
$$= \frac{\pi^2}{4\sin(2\pi\lambda)}\left[J_{-\lambda}(pe^{i\vartheta})J_{-\lambda}(pe^{-i\vartheta}) - J_{\lambda}(pe^{i\vartheta})J_{\lambda}(pe^{-i\vartheta})\right].$$

Combining the above discussion with the spectral decomposition of $\langle P_m(\cdot, s_1), P_n(\cdot, \overline{s_2})\rangle$, which is a direct consequence of (2.7.5), we obtain an identity that is exactly the same as the specialization of (2.7.7) with $h(r) = \pi\Theta(s_1 - \frac{1}{2}, s_2 - \frac{1}{2}; ir)|\Gamma(1+ir)|^{-2}$ (cf. (2.6.22)). By analytic continuation, the identity holds for $\mathrm{Re}\, s_1, \mathrm{Re}\, s_2 > \frac{3}{2}$. We then put $s_1 = s$, $s_2 = 2$, and proceed in much the same way as we did in the later part of the previous section. This ends the sketch of the proof of Theorem 2.6.

It should be noted that we actually needed a result on the spectral mean square of $\rho_j(n)$'s which is an analogue of (2.3.2). Thus we remark that the bound

$$\sum_{K/2 \le \kappa_j \le K} |\rho_j(n)|^2 e^{-\pi\kappa_j} \ll K^2 + |m|^2 K$$

is relatively easy to prove. This comes out from another transformation of (2.7.12): providing $|\mathrm{Re}\,(s_1 - s_2)| < 2\alpha$, $\mathrm{Re}\,(s_1 + s_2) > 2 + 4\alpha$, we have

$$B(p, \vartheta; s_1, s_2) = \frac{\Gamma(\frac{1}{2}(s_1 + s_2 - 1))}{8\pi^{\frac{3}{2}} i}$$
$$\times \int_0^1\int_0^1 u^{\frac{1}{2}(s_1-s_2)-1}(1-u)^{\frac{1}{2}(s_2-s_1)-1}v^{-\frac{1}{2}}(1-v)^{\frac{1}{2}(s_1+s_2)-2}$$
$$\times \int_{(\alpha)} \frac{\Gamma(\eta)}{\Gamma(\frac{1}{2}(s_1+s_2)-\eta)}\left(\frac{4p^{-2}u(1-u)v}{1-(2\sin\vartheta)^2 u(1-u)v}\right)^{\eta} d\eta du dv.$$

We put $s_1 = 2 + it$, $s_2 = 2 - it$, and follow the argument of the proof of (2.3.2).

2.8 Notes for Chapter 2

Sections 2.1–2.5

The main results of the present chapter are due to Kuznetsov ([35], [36]). Despite the fact that his original arguments were not completely rigorous, the trace formulas that were first stated in a rather obscure mimeographed preprint [35] had the potential to cause a major change in analytic number theory. The initial motivation of Kuznetsov was the problem of Linnik [40] about the possible cancellation among Kloosterman sums. Since the introduction of Ramanujan's Farey dissection argument, problems in number theory have frequently been reduced to the estimation of arithmetic trigonometrical sums. If the problem has a multiplicative structure, it is often the case that sums of Kloosterman sums with variable moduli turn up; and thus Linnik's problem becomes relevant. As an application of Theorem 2.3 Kuznetsov demonstrated such a cancellation for the first time: He deduced

$$\sum_{l \le L} \frac{1}{l} S(m,n;l) \ll L^{\frac{1}{6}} (\log L)^{\frac{1}{3}},$$

where the implied constant depends on $m, n \geq 1$ (see [36, Theorem 3] as well as Goldfeld and Sarnak [12]). This indicates that in some cases the algebraic bound (2.1.4) due to Weil can be superseded, though on average, by an analytical means – the trace formula (2.4.7). It is indeed hard to suppose that any algebraic arguments would ever be able to detect such a cancellation. Thus it appears natural to expect that the spectral method should play a major rôle in number theory. Incidentally Stepanov proved Weil's bound elementarily. We note also that the use of (2.1.4) is not mandatory in developing trace formulas. Any non-trivial bounds like $S(m,n;l) \ll l^{\theta}$ with a $\theta < 1$ are sufficient for our purpose (see Rankin [61, Section 5.4]).

Prior to Kuznetsov's investigation, Selberg [66] had studied Linnik's problem. He considered the analytic continuation of the Kloosterman-sum zeta-function $Z_{m,n}(s)$ defined by (2.4.1) in a general setting. For this purpose he employed the inner-product of two Poincaré series to which the unfolding method and the spectral expansion were applied. In view of Selberg's investigation one may conclude that estimates as good as Kuznetsov's are possible only for groups which share basic arithmetic characteristics with the full modular group. This indicates that the full modular group occupies a special place in the theory of automorphic functions.

Kuznetsov's argument is an infusion of ingenious manipulations of exotic formulas involving Bessel functions into Selberg's method. Our simplification (Motohashi [54]) lies mainly in that we have dispensed with heavy use of the theory of Bessel functions. This was made possible by Lemma 2.1 which replaces Kuznetsov's main lemma [36, p. 315]; the latter is in good accordance with the left side of Petersson's trace formula (2.2.9), though less flexible than our formula (2.1.2). The effect of the simplification is notable in our proof of the spectral estimate (2.3.2) which is originally due to Kuznetsov, and of fundamental importance in the theory of cusp-forms. One should observe that in formula (2.3.4) parameters are well separated, a feature not available in the corresponding formula of Kuznetsov [36, (4.50)]. This will be exploited more in Section 3.5. By refining the proof of Lemma 2.4 it is possible to replace (2.3.2) by Kuznetsov's asymptotic formula [36, (2.29)]:

$$\sum_{\kappa_j \le K} \frac{|\rho_j(m)|^2}{\cosh(\pi\kappa_j)} = \pi^{-2}K^2 + O\left(K \log K + Km^{\varepsilon} + m^{\frac{1}{2}+\varepsilon}\right),$$

where ε is an arbitrary positive constant, and the implied constant depends only on ε. Our Theorem 2.2 is precisely the same as Kuznetsov [36, Theorem 1], but our proof is simpler and shorter than his. In passing we remark that the trace formula (2.3.16) was obtained by Bruggeman [4] independently, though in a less refined form and a little later than Kuznetsov.

The spectral expansion (2.4.2) of the Kloosterman-sum zeta-function is due to Kuznetsov [36, (7.26)], which is a completion, in the case of the full modular group, of Selberg's investigation [66]. The above proof (see Motohashi [51]) is different from theirs and much shorter. Our Theorem 2.3 is essentially the same as Kuznetsov [36, Theorem 2]; the difference is in the condition (2.4.6), which is, however, immaterial to applications. Again, our proof of the trace formula (2.4.7) is different from Kuznetsov's. We proved (2.4.2) first and then used the Mellin transform to deduce (2.4.7), while Kuznetsov proved his trace formula directly via his counterpart ([36, (4.50)]) of our (2.3.4). In doing this he used an expansion of smooth functions into a series of the Neumann type, which is originally due to Titchmarsh and others, but which is, however, dispensed with in our argument. The opposite-sign case that is developed in Section 2.5 was first treated by Kuznetsov [35], although he did not include it in [36]. The original claims in [35] contain errors, which are corrected in the above. Our proof of the trace formula (2.5.6) is much simpler than Kuznetsov's. He used the inversion formula (2.6.1), while our proof depends only on some basic properties of K-Bessel functions. Viewing applications in retrospect, it appears to us that Kuznetsov's contribution is more in his sum formulas (2.4.7), (2.5.14) than in the trace formulas (2.3.16), (2.5.6).

The extensions of Kuznetsov's trace formulas to more general situations such as the congruence groups and cusp-forms of arbitrary multipliers are of no special difficulty; and our argument works, too, without any essential changes. See for example Proskurin [60] and Deshouillers and Iwaniec [8] modulo some blemishes, which are reworkings of Kuznetsov's argument, and can be simplified considerably if combined with our procedure. Also the cusp-forms of arbitrary weights can be treated as well. In the case of congruence subgroups we need to have bounds, similar to (2.1.4), for generalized Kloosterman sums which come from the Fourier expansion of Poincaré series at various cusps. A thorough treatment of these important trigonometrical sums is developed in Rankin [61].

What we said about holomorphic cusp-forms in the second section is obviously very basic. The precise formula for the dimension of the space of holomorphic cusp-forms with given weight over an arbitrary Fuchsian group is usually proved by using the Riemann–Roch theorem. But the particular assertion (2.2.4) can be proved elementarily; see Rankin [61, p. 198]. However, we avoided the use of (2.2.4) in the proof of Petersson's trace formula (2.2.9). This is because we wanted to make our argument susceptible of extensions to, e.g., congruence subgroups. The use of the convergence parameter s in (2.2.13) is an idea of Hecke [17, p. 391], which brings the situation close to the theory of real-analytic Poincaré series.

No account of the spectral resolution of Δ would be complete without mentioning Selberg's trace formula and his zeta-function. It appears, however, that they do not have any immediate relevance to our applications of the spectral theory to the Riemann zeta-function, at least presently. Thus readers are referred to other sources.

Sections 2.6–2.7

This part of the chapter is an exotic excursion. It does not have much to do with the later part of our monograph. It is, however, worth having a wider perspective in order to see the whole story structurally.

What we intended to show in Section 2.6 is the informal diagram:

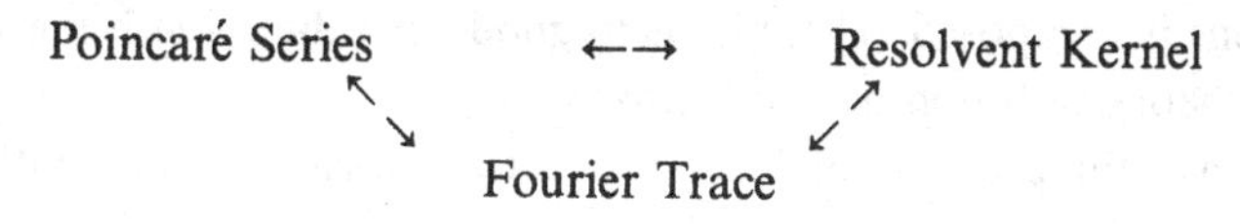

The bottom line stands for the device of equating Fourier coefficients of an arithmetically defined automorphic function; they are picked up from either the definition itself or the spectral expansion of the function. The idea was first harnessed by Selberg in his seminal work [66], and later developed by Kuznetsov [35] and many others. The arrows on the left indicate what was dealt with in Sections 2.1–2.5; thus Poincaré series are used to pick up particular Maass–Fourier coefficients. The arrows on the right are essentially due to Fay [11]. More precisely this is the fact that his functional equation [11, Corollary 3.6] for

$$Z_s(m,n) = \sum_{l=1}^{\infty} \frac{1}{l} S(m,n;l) \tilde{J}_{2s-1}\left(\frac{4\pi}{l}\sqrt{|mn|}\right),$$

where $\tilde{J}_\nu$ is as in (2.6.22), is equivalent to the trace formulas (2.3.16) and (2.5.6), which has been observed by several people independently (see, e.g., Iwaniec [27, Chapter 9] for details). The function $Z_\alpha(m,n)$ appears in the Fourier expansion of the resolvent kernel $R_\alpha(z,w)$, the first proof of which is due to Niebur [58] and Fay [11] though the latter is worked out in a far more general situation (cf. Iwaniec [27, Chapter 5]). Their argument depends on a difficult analysis of some differential equations related to Bessel functions. So it might be worth remarking that an extension of our investigation given in Section 2.6 seems to yield a simple proof of the Fourier expansion of $R_\alpha(z,w)$. In fact it is possible to deduce a new Fourier expansion of $P_m(z,s)$ by applying the Mellin inversion to the relations (2.6.14) and (2.6.21). Then we may use this expansion to pick up Fourier coefficients of the $R_\alpha(z,w)$. Thus it appears reasonable to suppose that the integral formulas in Lemma 2.7 are related to a fundamental symmetry of the non-Euclidean plane. The top line in the above diagram indicates this observation, though it is partly conjectural. In this context we stress also that the Niebur–Fay expansion of $R_\alpha(z,w)$ is a consequence of the trace formulas (2.3.16) and (2.5.6), although the actual procedure is to apply them to the iterated kernel and take a certain limit to reach the original kernel. Since it is natural to suppose that all the real-analytic structure induced by a given Fuchsian group is, in principle, contained in the corresponding resolvent of the non-Euclidean Laplacian, the above diagram implies that real-analytic Poincaré series constitute the basic ingredient of which the entire space of automorphic functions is composed. This fact is in good accordance with the situation in the theory of holomorphic forms.

We gave three proofs of Kuznetsov's trace formulas. The author likes best the third proof given in the later part of Section 2.6. For it is the most general and structural among the three, as can be seen also from its extension that is developed in Section 2.7. The reduction of the width of the horizontal strip pertaining to the weight function $h(r)$ is due to Yoshida [76]. Like us he started from our (2.6.22) but he proceeded in a different way; his is related to Niebur [58]. Our use of (2.6.25) made the reasoning simpler and more direct. It should be stressed that an equivalent observation was made by us in [57] quite independently. In fact, the application of the formula (2.6.25) was first performed in [57], where a weight function similar to the above occurs. We needed it to be regular in the strip given in Theorem 2.6. If translated to the case of the full modular group, this corresponds exactly to (2.6.24).

In this context it is worth mentioning that in treating the Kloosterman-sum part the decay condition in (2.6.24) is too stringent; in fact the bound $h(r) \ll (1 + |r|)^{-1-\delta}$ is sufficient. This suggests to us that there should be a version of trace formulas in which the spectral part is summed in a new way still to be discovered and the weight function decays relatively slowly.

Here are some minor remarks: The formula (2.6.1) is known as the Kontorovitch–Lebedev inversion formula. For a more precise formulation see Lebedev [38, (5.14.14)]. At (2.6.4) the basic theory of resolvents of ordinary differential equations of second order is used. For this see, e.g., Courant and Hilbert [7, Chapter V]. The formula (2.6.7) is given in Gradshteyn and Ryzhik [15, 6.654]; also (2.6.8) is listed there as the formula 6.794(9). The latter contains, however, a blemish. As to Barnes' integral formula (2.6.13) and related topics see either Titchmarsh [69] or Whittaker and Watson [75].

What is given in Section 2.7 is a part of our recent investigation (Motohashi [57]). The similarity between (2.3.16) and (2.7.7) is remarkable. The upper half space model of the Lobachevsky geometry was implemented by Beltrami, as was mentioned before; and the first example of its discontinuous motion groups was given by Picard, who introduced the fundamental domain (2.7.4). The argument of Section 2.7 is naturally not limited to the Picard group. It is, in fact, a matter of technicality to extend it to Bianchi groups defined over arbitrary imaginary quadratic number fields, although we shall have additional complexity caused by the existence of inequivalent cusps, the number of which is equal to the class number of the underlying field. It should be mentioned that the spectral theory of $\mathcal{H}^3$ is briefly indicated in Selberg [65, p. 79].

We provide here some clues to our condensed argument: The assertion (2.7.6) is proved by solving the differential equation

$$\left[(\varrho^2+\varrho)(d/d\varrho)^2+3(\varrho+\tfrac{1}{2})(d/d\varrho)+\alpha(2-\alpha)\right]f(\varrho)=0,$$

which is an analogue of (1.2.7), and equivalent to $(\Delta+\alpha(2-\alpha))_w f(\varrho(z,w))=0$ in the new context. The bound (2.7.10) can be proved in much the same way as (2.1.4). The Mellin transform (2.7.11) is given in [15, 6.621] and [74, p. 385]. The formula (2.7.13) is due to Sonine [68] (see also [15, 6.596] and [74, p. 417]). The elegant formula (2.7.15) is tabulated as [15, 6.688(1)], and can probably be attributed to Gegenbauer. Since its proof is hard to find in the literature, we show the outline: We expand the integrand by using the defining series for the J-Bessel and the cosine functions, and integrate termwise. An elementary rearrangement of the resulting double series shows that the original integral is equal to

$$\frac{\sqrt{\pi}(b/2)^{2\lambda}}{2\Gamma(2\lambda+\frac{1}{2})}\int_0^1\frac{J_\lambda(\sqrt{a^2+b^2u})}{(a^2+b^2u)^\lambda}(u(1-u))^{\lambda-\frac{1}{2}}du.$$

Or we may start from this; insertion of the series expansion for J_λ and termwise integration readily yield the original integral. Putting $u=(\sin\theta)^2$ in this expression we see that it reduces to the Gegenbauer integral (see [15, 6.684(1)] and [74, p. 367]; a rigorous treatment is given in [68, p. 37]). This ends the proof. Alternatively, we may proceed as follows: It is easier to deal with the equivalent formula ([15, 6.688(3)]):

$$\int_0^{\frac{1}{2}\pi}\cos((\alpha-\beta)\cos\tau)J_{2\lambda}(2\sqrt{\alpha\beta}\sin\tau)d\tau=\frac{1}{2}\pi J_\lambda(\alpha)J_\lambda(\beta),$$

where $\lambda>-\frac{1}{2}$, and $\alpha,\beta>0$. We differentiate the left side twice with respect to α, and rearrange the result using the differential equation satisfied by $J_{2\lambda}$. Then we see that the integral is a solution of the differential equation for J_λ. Hence we find that it is a constant multiple of $J_\lambda(\alpha)$, for the integral is obviously of order $O(\alpha^\lambda)$ as α tends to 0. The computation of this constant is immediate.

It may be fascinating that a relation between these integral formulas and some representations of elementary Lie groups is observable in our trace formula (2.7.7) (cf. Vilenkin [71]). This is, however, by no means an unexpected phenomenon. For we are actually studying the analytic structure of geometrical spaces which are in fact quotients of certain Lie groups.

3
Automorphic L-functions

In this chapter we shall collect some basic facts about L-functions derived from cusp-forms over the full modular group Γ by way of the Hecke correspondence. We shall see in the next chapter that these L-functions appear as components of our explicit formula for the fourth power moment of the Riemann zeta-function. This fact is in no sense superficial: In the course of the proof of the mean value we shall have an expression involving Kloosterman sums, which is spectrally decomposed by means of Theorems 2.3 and 2.5. But this will be made initially only in a domain of relevant parameters which does not contain the point we are most interested in. Thus we shall face the problem of analytic continuation; and its solution will depend indispensably on the analytical properties of these L-functions. In this process of analytic continuation the multiplicative property of Hecke operators will play an important rôle. Hence we shall develop the essentials of Hecke's idea in the first section. Once the analytic continuation is completed and the explicit formula for the mean value is established, we shall need good spectro-statistical estimates of special values of L-functions to extract quantitative information from the formula. Having this application in mind we shall give an account of spectral mean values of Hecke L-functions in the later sections of this chapter.

3.1 Hecke operators

In order to define Hecke operators we introduce, for a given positive integer n,

$$\mathbb{M}(n) = \left\{ z \mapsto \frac{az+b}{lz+h} : a, b, l, h \in \mathbb{Z} \text{ with } ah - bl = n \right\},$$

and consider the transforms

$$f(M(z)) \quad (M \in \mathbb{M}(n))$$

of an arbitrary Γ-automorphic function f. Since all members in the set $\Gamma M \subseteq \mathbb{M}(n)$ yield the same function $f(M(z))$, this is equivalent to classifying the elements of $\mathbb{M}(n)$ according to the left Γ-cosets. Let $\{M_\nu\}$ be a representative set in this classification so that

$$\mathbb{M}(n) = \bigcup_\nu \Gamma M_\nu \quad \text{(disjoint)}.$$

For each $\gamma \in \Gamma$ the set $\{M_\nu \gamma\}$ is obviously a permutation mod Γ of the original set; and thus the function

$$\sum_\nu f(M_\nu(z)) \tag{3.1.1}$$

is again Γ-automorphic, provided the sum is finite. We shall show that

$$\{z \mapsto (az+b)/d \,:\, ad = n,\ 0 \le b < d\} \tag{3.1.2}$$

is such a representative set; in particular the sum (3.1.1) is well-defined. In fact, it is easy to see that for any $M \in \mathbb{M}(n)$ there is a $\gamma \in \Gamma$ such that the denominator of $(\gamma M)(z)$ is a constant; and then taking into account the action of Γ_∞ (see (1.1.4)) we find that ΓM contains an element belonging to the set (3.1.2). On the other hand, that the elements in (3.1.2) are mutually inequivalent mod Γ is immediate.

Hence

$$(T(n)f)(z) = n^{-\frac{1}{2}} \sum_{ad=n} \sum_{b \bmod d} f((az+b)/d) \tag{3.1.3}$$

is Γ-automorphic. This $T(n)$ is called the Hecke operator acting over the linear space of all Γ-automorphic functions. We then have the fundamental

Lemma 3.1 *Hecke operators commute with each other, satisfying, for any* $m, n \ge 1$,

$$T(m)T(n) = \sum_{d|(m,n)} T\left(\frac{mn}{d^2}\right). \tag{3.1.4}$$

Proof We suppose first that $(m,n) = 1$. Then we have, for any Γ-

automorphic f,

$$\begin{aligned}(T(m)T(n)f)(z) &= (mn)^{-\frac{1}{2}} \sum_{ad=m} \sum_{b \bmod d} \sum_{a'd'=n} \sum_{b' \bmod d'} f((aa'z + a'b + b'd)/(dd')) \\ &= (mn)^{-\frac{1}{2}} \sum_{ad=m} \sum_{a'd'=n} \sum_{h \bmod dd'} f((aa'z + h)/(dd')) = (T(mn)f)(z),\end{aligned}$$

since $(a', d) = 1$. This is equivalent to (3.1.4) in the present case.

Thus we may restrict ourselves to the situation where m, n are powers of a given prime p: We have to show, for any $u, v \geq 0$,

$$T(p^u)T(p^v) = \sum_{l=0}^{\min(u,v)} T(p^{u+v-2l}). \tag{3.1.5}$$

For this purpose we put

$$(T(p)f)(z) = p^{-\frac{1}{2}}\Big\{f(pz) + \sum_{h=0}^{p-1} f((z+h)/p)\Big\}. \tag{3.1.6}$$

Then we have

$$\begin{aligned}&(T(p^r)T(p)f)(z) = p^{-(r+1)/2} \\ &\times \sum_{u=0}^{r} \sum_{b=0}^{p^u-1} \Big\{f((p^{r-u+1}z + pb)/p^u) + \sum_{h=0}^{p-1} f((p^{r-u}z + b + hp^u)/p^{u+1})\Big\}.\end{aligned}$$

In this the numbers $b + hp^u$ are incongruent to each other $\bmod p^{u+1}$; thus the terms involving h contribute, in total, $(T(p^{r+1})f)(z) - p^{-(r+1)/2}f(p^{r+1}z)$. The remaining part is divided into two: If $u = 0$, then we get only one term $f(p^{r+1}z)$. If $u \geq 1$, then the sum over b is equal to

$$p \sum_{b'=0}^{p^{u-1}-1} f((p^{r-u}z + b')/p^{u-1})$$

because of the periodicity of f; and the contribution is, in total, $(T(p^{r-1})f)(z)$. Altogether, we get

$$T(p^r)T(p) = T(p^{r+1}) + T(p^{r-1}) \quad (r \geq 1), \tag{3.1.7}$$

which is a special case of (3.1.5). This shows, in particular, that $T(p^r)$ is a polynomial in $T(p)$; and we have proved the first assertion of the lemma.

We then put

$$U(\lambda,\mu) = T(p^\lambda)T(p^\mu) - T(p^{\lambda+1})T(p^{\mu-1}) \quad (\lambda \geq 0,\ \mu \geq 1).$$

We modify this by $T(p^{\lambda+1}) = T(p^\lambda)T(p) - T(p^{\lambda-1})$ which is (3.1.7) with $r = \lambda$; then use (3.1.7) again but with $r = \mu - 1$. We get

$$U(\lambda,\mu) = U(\lambda-1,\mu-1)$$

provided $\mu \geq 2$. Thus we have, for $1 \leq l \leq u \leq v$,

$$U(u+v-l,l) = U(u+v-2l+1,1) = T(p^{u+v-2l}).$$

Summing this for $l = 1,2,\ldots,u$ we get (3.1.5), and end the proof of the lemma.

Next, we consider Hecke operators in the framework of the metric theory:

Lemma 3.2 *Hecke operators are Hermitian at each eigen-space of* Δ *in* $L^2(\mathcal{F}, d\mu)$.

Proof Since $\mathbb{M}(n) \subset \mathbb{T}(\mathcal{H})$, the operator Δ commutes with Hecke operators (see Section 1.1). Because of this and what we have seen in the proof of the previous lemma, it is sufficient to prove that for a given prime p and any $f_1, f_2 \in L^2(\mathcal{F}, d\mu)$ we have

$$\langle T(p)f_1, f_2\rangle = \langle f_1, T(p)f_2\rangle. \tag{3.1.8}$$

To this end we rewrite (3.1.6) as

$$(T(p)f)(z) = p^{-\frac{1}{2}}\sum_{h=0}^{p} f(\iota_p\gamma_h(z)),$$

where

$$\iota_p(z) = -1/(pz); \quad \gamma_h(z) = -1/(z+h) \quad (0 \leq h < p), \quad \gamma_p(z) = z.$$

Thus (3.1.8) is equivalent to the identity

$$\begin{aligned}\int_{\mathcal{P}} f_1(\iota_p(z))\overline{f_2(z)}d\mu(z) &= \int_{\mathcal{P}} f_1(z)\overline{f_2(\iota_p(z))}d\mu(z)\\ &= \int_{\iota_p(\mathcal{P})} f_1(\iota_p(z))\overline{f_2(z)}d\mu(z),\end{aligned} \tag{3.1.9}$$

where

$$\mathcal{P} = \bigcup_{h=0}^{p} \gamma_h(\mathcal{F}). \tag{3.1.10}$$

To prove (3.1.9) we need to introduce the congruence subgroup

$$\Gamma_0(p) = \left\{ z \mapsto \frac{az+b}{lz+h} : a, b, l, h \in \mathbb{Z} \text{ with } ah - bl = 1,\, l \equiv 0 \bmod p \right\}. \tag{3.1.11}$$

Then we shall show that

$$\Gamma = \bigcup_{h=0}^{p} \Gamma_0(p)\gamma_h \quad \text{(disjoint).} \tag{3.1.12}$$

To this end let a and c be the upper left and the lower left entries, respectively, of a pull-back to $SL(2, \mathbb{Z})$ of a given $\gamma \in \Gamma$. If c is divisible by p then obviously γ is in the coset $\Gamma_0(p)\gamma_p$. Otherwise there exist integers α, β such that $ap\alpha - c\beta = 1$. This is the same as that the coset $\Gamma_0(p)\gamma$ contains an element whose pull-back to $SL(2, \mathbb{Z})$ has the upper left entry equal to 0. Taking into account the action of the maps $z \mapsto z/(prz + 1)$ ($r \in \mathbb{Z}$), we conclude that γ is in one of $\Gamma_0(p)\gamma_h$ $(0 \le h < p)$. On the other hand it is easy to see that γ_h $(0 \le h \le p)$ are inequivalent $\bmod\, \Gamma_0(p)$. Thus we get (3.1.12).

In particular we see that $\mathcal{P}$ is a fundamental domain of $\Gamma_0(p)$; that is, the family of domains $\{\tau(\mathcal{P}),\, \tau \in \Gamma_0(p)\}$ induces a tessellation of the upper half plane. Then we observe that $\iota_p(\mathcal{P})$ is also a fundamental domain of $\Gamma_0(p)$. This is due to the Fricke identity:

$$\iota_p \Gamma_0(p) = \Gamma_0(p)\iota_p,$$

which is easy to verify. This relation yields also that $f_1(\iota_p(z))$ in (3.1.9) is $\Gamma_0(p)$-automorphic. Hence our problem has been reduced to showing that for an arbitrary $\Gamma_0(p)$-automorphic f we have

$$\int_{\mathcal{P}} f(z)d\mu(z) = \int_{\iota_p(\mathcal{P})} f(z)d\mu(z)$$

whenever both sides converge absolutely. But this is straightforward; and we end the proof of the lemma.

Because of the last two lemmas and a basic assertion on commuting Hermitian operators, we conclude that the orthonormal system $\{\psi_j(z) : j \ge 1\}$ introduced at (1.1.43) can be chosen in such a way that we have, for each pair j, $n \ge 1$,

$$T(n)\psi_j = t_j(n)\psi_j \tag{3.1.13}$$

with certain *real* numbers $t_j(n)$. We call these numbers Hecke eigenvalues. The formula (3.1.4) implies

$$t_j(m)t_j(n) = \sum_{d\,|\,(m,n)} t_j\left(\frac{mn}{d^2}\right). \tag{3.1.14}$$

We may also take into account the action of the reflection operator $z \mapsto -\bar{z}$, which is obviously an involution on the space $L^2(\mathcal{F}, d\mu)$, and commutes with all Hecke operators. Thus the system can be assumed to satisfy, besides (3.1.13),

$$\psi_j(-\bar{z}) = \varepsilon_j \psi_j(z) \quad (\varepsilon_j = \pm 1) \tag{3.1.15}$$

for any $j \geq 1$. We shall assume (3.1.13) and (3.1.15) throughout the remainder of this monograph.

In terms of Maass–Fourier coefficients (see (1.1.41)) these are equivalent to the relations

$$\rho_j(n) = \rho_j(1)t_j(n), \quad \rho_j(-n) = \varepsilon_j \rho_j(n) \quad (j,\, n \geq 1). \tag{3.1.16}$$

For we have, by definition,

$$\begin{aligned}(T(n)\psi_j)(z) &= n^{-\frac{1}{2}} \sum_{ad=n} \sqrt{ay/d} \sum_{m\neq 0} \rho_j(m) K_{i\kappa_j}(2\pi a|m|y/d) e(amx/d) \\ &\quad \times \sum_{b \bmod d} e(bm/d) \\ &= \sqrt{y} \sum_{ad=n} \sum_{m\neq 0} \rho_j(dm) K_{i\kappa_j}(2\pi a|m|y) e(amx).\end{aligned}$$

Then (3.1.13) is the same as having

$$t_j(n)\rho_j(m) = \sum_{d\,|\,(m,n)} \rho_j\left(\frac{mn}{d^2}\right).$$

This gives, in particular, the first relation in (3.1.16), which in turn yields this identity in view of (3.1.14). Since the second relation in (3.1.16) is trivial, we have confirmed the above claim. In passing we note that we have, for all $j \geq 1$,

$$\rho_j(1) \neq 0. \tag{3.1.17}$$

For otherwise we would have $\psi_j \equiv 0$.

On the bound for Hecke eigenvalues we have

Lemma 3.3 *For all* $j, n \geq 1$

$$|t_j(n)| \ll n^{\frac{1}{4}+\delta}, \tag{3.1.18}$$

where δ is an arbitrary fixed positive constant, and the implied constant depends only on δ.

Proof This is to use the power of multiplicativity: Let us assume first that we have the bound

$$|\rho_j(1)||t_j(n)| \le \tau n^\alpha \quad (\alpha > 0)$$

uniformly in n, where τ is independent of n; note that the assertion (2.3.2) implies that any $\alpha > \frac{1}{4}$ is admissible. Inserting this into the right side of (3.1.14) with $m = n$ we have

$$|\rho_j(1)||t_j(n)|^2 \le \tau\sigma_{-2\alpha}(n)n^{2\alpha};$$

or

$$|\rho_j(1)||t_j(n)| \le \tau^{\frac{1}{2}}|\rho_j(1)\xi_{2\alpha}(n)|^{\frac{1}{2}}n^\alpha,$$

where

$$\xi_\sigma(n) = \prod_{p|n}(1-p^{-\sigma})^{-1} \quad (p: \text{ prime}).$$

Repeating the same procedure, we get, for any $r \ge 1$,

$$|\rho_j(1)||t_j(n)| \le \tau^{2^{-r}}|\rho_j(1)\xi_{2\alpha}(n)|^{1-2^{-r}}n^\alpha.$$

Taking r to infinity we get

$$|t_j(n)| \le \xi_{2\alpha}(n)n^\alpha,$$

which ends the proof.

Next, turning to the space $\mathscr{C}_k(\Gamma)$ of holomorphic cusp-forms of weight $2k$ (see Section 2.2), we define the Hecke operator $T_k(n)$ by

$$(T_k(n)f)(z) = n^{-\frac{1}{2}}\sum_{ad=n}(a/d)^k\sum_{b=1}^{d} f((az+b)/d). \tag{3.1.19}$$

The right side belongs to $\mathscr{C}_k(\Gamma)$ if f does so. To see this, we rewrite the above as

$$(T_k(n)f)(z) = n^{-\frac{1}{2}}\sum_{\nu} f(M_\nu(z))\Big(\frac{d}{dz}M_\nu(z)\Big)^k,$$

where M_ν runs over the representative set (3.1.2). Then we have, for an arbitrary $\gamma \in \Gamma$,

$$(T_k(n)f)(\gamma(z))j(\gamma,z)^{-2k} = n^{-\frac{1}{2}} \sum_{\nu} f((M_\nu\gamma)(z)) \Big(\frac{d}{dz}(M_\nu\gamma)(z)\Big)^k$$
$$= n^{-\frac{1}{2}} \sum_{\nu} f((\tau_\nu M_\nu)(z)) \Big(\frac{d}{dz}(\tau_\nu M_\nu)(z)\Big)^k$$

with certain τ_ν's in Γ. This gives $(T_k(n)f)(\gamma(z)) = (T_k(n)f)(z)j(\gamma,z)^{2k}$, for $f(\tau(z))(d\tau(z)/dz)^k = f(z)$ if $\tau \in \Gamma$. Since $(T_k(n)f)(z)$ obviously vanishes at the cusp of Γ, we get the above assertion.

Now the argument developed for the Hecke operators $T(n)$ applies to $T_k(n)$ as well. We have

$$T_k(m)T_k(n) = \sum_{d\,|\,(m,n)} T_k\Big(\frac{mn}{d^2}\Big)$$

as well as

$$\langle T_k(n)f_1, f_2\rangle_k = \langle f_1, T_k(n)f_2\rangle_k \quad \big(f_1, f_2 \in \mathscr{C}_k(\Gamma)\big)$$

(see (2.2.1)). In particular we can assume that the system $\{\psi_{j,k} : 1 \le j \le \vartheta(k)\}$ introduced at (2.2.2) is such that for all $n \ge 1$

$$T_k(n)\psi_{j,k} = t_{j,k}(n)\psi_{j,k} \tag{3.1.20}$$

with certain real numbers $t_{j,k}(n)$, which are again called Hecke eigenvalues. Thus we have

$$\rho_{j,k}(n) = \rho_{j,k}(1)t_{j,k}(n) \tag{3.1.21}$$

in the expansion (2.2.3). Further, we have

$$t_{j,k}(n) \ll n^{\frac{1}{4}+\delta} \tag{3.1.22}$$

for any fixed $\delta > 0$ by virtue of (2.2.10), where the implied constant depends only on δ. We shall assume (3.1.20)–(3.1.22) throughout the remainder of this monograph.

3.2 *L*-functions

Now we introduce a Dirichlet series attached to the cusp-form ψ_j:

$$L_j(s) = \sum_{n=1}^{\infty} \rho_j(n)n^{-s}. \tag{3.2.1}$$

This is called either the Hecke L- or the automorphic L-function derived from ψ_j. Providing ψ_j satisfies (3.1.13), we put

$$H_j(s) = \sum_{n=1}^{\infty} t_j(n) n^{-s}, \tag{3.2.2}$$

which is also called the Hecke L-function attached to ψ_j. Obviously we have

$$L_j(s) = \rho_j(1) H_j(s).$$

Also we introduce the Rankin L-function attached to a pair of cusp-forms $\psi_j, \psi_{j'}$:

$$R_{j,j'}(s) = \sum_{n=1}^{\infty} \rho_j(n)\overline{\rho_{j'}(n)} n^{-s}. \tag{3.2.3}$$

These are obviously regular for $\mathrm{Re}\, s > \frac{3}{2}$, say, because of the bound (2.3.2) or (3.1.18) (see Section 3.6).

We have more precisely

Lemma 3.4 *The function $H_j(s)$ is entire, and satisfies the functional equation*

$$H_j(s) = 2^{2s-1}\pi^{2(s-1)}\Gamma(1-s+i\kappa_j)\Gamma(1-s-i\kappa_j) \times \{\varepsilon_j \cosh(\pi\kappa_j) - \cos(\pi s)\} H_j(1-s). \tag{3.2.4}$$

In particular we have

$$H_j(s) \ll \kappa_j^c \tag{3.2.5}$$

uniformly for $j \geq 1$ and bounded s, where c depends only on $\mathrm{Re}\, s$. *Also we have, in the region of absolute convergence,*

$$H_j(s) = \prod_p \left(1 - t_j(p)p^{-s} + p^{-2s}\right)^{-1} \tag{3.2.6}$$

as well as

$$H_j(s)H_j(s-\alpha)/\zeta(2s-\alpha) = \sum_{n=1}^{\infty} \sigma_\alpha(n) t_j(n) n^{-s}. \tag{3.2.7}$$

Proof The bound (3.2.5) is a simple consequence of the combination of (3.1.18), (3.2.4), Stirling's formula, and the Phragmén–Lindelöf convexity principle. The Euler product representation (3.2.6) is an easy consequence of the relations $t_j(mn) = t_j(m)t_j(n)$ for $(m,n) = 1$ and $t_j(p)t_j(p^r) =$

$t_j(p^{r+1}) + t_j(p^{r-1})$ $(r \geq 1)$. Further, the assertion (3.2.7) is equivalent to the Möbius inversion of (3.1.14):

$$t_j(mn) = \sum_{d \mid (m,n)} \mu(d) t_j\left(\frac{m}{d}\right) t_j\left(\frac{n}{d}\right). \tag{3.2.8}$$

Thus we need only to prove the functional equation.

We first deal with the case $\varepsilon_j = 1$. We consider the integral

$$L_j^+(s) = \int_0^\infty \psi_j(iy) y^{s-\frac{3}{2}} dy,$$

where $\operatorname{Re} s$ is supposed to be sufficiently large. Since $\psi_j(iy) = \psi_j(i/y)$, $y > 0$, we have

$$L_j^+(s) = \int_1^\infty \psi_j(iy)\{y^{s-\frac{3}{2}} + y^{-\frac{1}{2}-s}\} dy.$$

This converges for any s because $\psi_j(iy)$ is of exponential decay when y increases. Thus $L_j^+(s)$ is entire, and satisfies

$$L_j^+(s) = L_j^+(1-s).$$

On the other hand we have, by (1.1.41) and (3.1.16),

$$L_j^+(s) = 2\rho_j(1) \sum_{n=1}^\infty t_j(n) \int_0^\infty K_{i\kappa_j}(2\pi n y) y^{s-1} dy.$$

The exchange of the order of the sum and the integral has to be verified, but it is enough to observe that (1.1.17) gives $|K_{ir}(x)| \leq K_0(x)$ for real r and $x > 0$. Hence we have, by (2.6.12),

$$L_j^+(s) = \frac{1}{2}\rho_j(1)\pi^{-s}\Gamma(\tfrac{1}{2}(s + i\kappa_j))\Gamma(\tfrac{1}{2}(s - i\kappa_j))H_j(s),$$

which implies that $H_j(s)$ is entire and

$$H_j(s) = \pi^{2s-1} \frac{\Gamma(\frac{1}{2}(1 - s + i\kappa_j))\Gamma(\frac{1}{2}(1 - s - i\kappa_j))}{\Gamma(\frac{1}{2}(s + i\kappa_j))\Gamma(\frac{1}{2}(s - i\kappa_j))} H_j(1-s).$$

The right side is simplified by the duplication formula for the Γ-function; and we get (3.2.4) with $\varepsilon_j = 1$.

We consider next the case $\varepsilon_j = -1$. For this purpose we introduce

$$\psi_j^-(z) = y(\partial/\partial x - i\partial/\partial y)\psi_j(z). \tag{3.2.9}$$

The Γ-invariance of ψ_j and the Cauchy–Riemann relation for the function $\gamma(z)$ imply that ψ_j^- is of weight 2, that is,

$$\psi_j^-(\gamma(z)) = \psi_j^-(z)(j(\gamma, z)/|j(\gamma, z)|)^2 \quad (\gamma \in \Gamma). \tag{3.2.10}$$

We then put

$$L_j^-(s) = \int_0^\infty \psi_j^-(iy) y^{s-\frac{3}{2}} dy.$$

The relation $\psi_j(z) = \psi_j(-1/z)$ gives $\psi_j^-(i/y) = -\psi_j^-(iy)$, $y > 0$; and thus

$$L_j^-(s) = \int_1^\infty \psi_j^-(iy)\{y^{s-\frac{3}{2}} - y^{-\frac{1}{2}-s}\}dy.$$

Hence $L_j^-(s)$ is entire and satisfies $L_j^-(s) = -L_j^-(1-s)$. On the other hand we have

$$\psi_j^-(iy) = 4\pi i y^{\frac{3}{2}} \sum_{n=1}^\infty n\rho_j(n) K_{i\kappa_j}(2\pi n y) - iy(\partial/\partial y)\psi_j(iy),$$

where the last factor vanishes, for $\psi_j(z) \equiv 0$ on the imaginary axis if $\varepsilon_j = -1$. This gives

$$L_j^-(s) = i\rho_j(1)\pi^{-s}\Gamma(\tfrac{1}{2}(1+s+i\kappa_j))\Gamma(\tfrac{1}{2}(1+s-i\kappa_j))H_j(s);$$

and we end the proof of the lemma.

We now move to Rankin L-functions. Their basic analytical properties are to be used in Section 3.5. These are embodied in

Lemma 3.5 *If the cusp-forms ψ_j, $\psi_{j'}$ satisfy* (3.1.15), *then the function $R_{j,j'}(s)$ continues meromorphically to the whole of $\mathbb{C}$. If $j \neq j'$ then it is regular for* $\mathrm{Re}\, s \geq \frac{1}{2}$. *If $j = j'$ then it is regular for* $\mathrm{Re}\, s \geq \frac{1}{2}$ *save for the simple pole at $s = 1$ with the residue $12\pi^{-2}\cosh(\pi\kappa_j)$. We have also the functional equation*

$$R^*_{j,j'}(s) = R^*_{j,j'}(1-s), \tag{3.2.11}$$

where

$$R^*_{j,j'}(s) = (2\pi^2)^{-s}\zeta(2s)\Gamma(s+\tfrac{1}{2}(1-\varepsilon_j\varepsilon_{j'})) \times \Gamma(s+\tfrac{1}{2}(1-\varepsilon_j\varepsilon_{j'}); i\kappa_j, i\kappa_{j'})R_{j,j'}(s)$$

with

$$\Gamma(s;u,v) = \Gamma(\tfrac{1}{2}(s+u+v))\Gamma(\tfrac{1}{2}(s+u-v)) \times \Gamma(\tfrac{1}{2}(s-u+v))\Gamma(\tfrac{1}{2}(s-u-v))\Gamma(s)^{-1}. \tag{3.2.12}$$

In particular, if ψ_j, $\psi_{j'}$ satisfy (3.1.13) *and* (3.1.15), *then we have*

$$R_{j,j'}(s) \ll |\rho_j(1)\rho_{j'}(1)|(\kappa_j\kappa_{j'}|s|)^c \tag{3.2.13}$$

provided s is well off the point 1, and stays in a fixed vertical strip in the *half plane* $\mathrm{Re}\, s \geq \frac{1}{2}$. *Here the exponent* c *and the implied constant are absolute.*

Proof For the sake of notational simplicity we put, within this proof,

$$\kappa_j = \kappa, \quad \kappa_{j'} = \kappa'.$$

We consider first the case $\varepsilon_j \varepsilon_{j'} = +1$. We put

$$V_{j,j'}(s) = \int_{\mathscr{F}} \psi_j(z)\overline{\psi_{j'}(z)} E(z,s) d\mu(z), \tag{3.2.14}$$

where the notation is as in the second chapter. Here the integrand decays exponentially as z tends to the cusp. By Lemma 1.2 the singularities of $E(z,s)$, which are all poles, do not depend on z but solely on the location of s; thus $V_{j,j'}(s)$ is meromorphic for all s. In particular $V_{j,j'}(s)$ is regular for $\mathrm{Re}\, s \geq \frac{1}{2}$ except for the possible simple pole at $s = 1$ with the residue

$$\int_{\mathscr{F}} \{\text{the residue of } E(z,s) \text{ at } s = 1\} \times \psi_j(z)\overline{\psi_{j'}(z)} d\mu(z) = 3\pi^{-1}\delta_{j,j'}. \tag{3.2.15}$$

Further, (1.1.11) gives the functional equation

$$\pi^{-s}\Gamma(s)\zeta(2s)V_{j,j'}(s) = \pi^{s-1}\Gamma(1-s)\zeta(2(1-s))V_{j,j'}(1-s). \tag{3.2.16}$$

On the other hand the unfolding method described at (2.1.7) yields, for $\mathrm{Re}\, s > 1$,

$$V_{j,j'}(s) = \int_0^\infty \int_{-\frac{1}{2}}^{\frac{1}{2}} \psi_j(x+iy)\overline{\psi_{j'}(x+iy)} y^{s-2} dx dy. \tag{3.2.17}$$

By (1.1.41) and the second relation in (3.1.16) we have

$$\begin{aligned} V_{j,j'}(s) &= 2\sum_{n=1}^{\infty} \rho_j(n)\overline{\rho_{j'}(n)} \int_0^\infty K_{i\kappa}(2\pi ny)K_{i\kappa'}(2\pi ny)y^{s-1}dy \\ &= 2(2\pi)^{-s}R_{j,j'}(s) \int_0^\infty K_{i\kappa}(y)K_{i\kappa'}(y)y^{s-1}dy. \end{aligned}$$

We have used the fact that $K_{i\nu}(y)$ is real for real ν as can be seen from (1.1.17). Thus the formula (2.6.11) gives immediately that for $\mathrm{Re}\, s > 1$

$$V_{j,j'}(s) = \tfrac{1}{4}\pi^{-s}\Gamma(s; i\kappa, i\kappa')R_{j,j'}(s). \tag{3.2.18}$$

This and (3.2.15), (3.2.16) give the assertions, save for (3.2.13), of the lemma provided $\varepsilon_j \varepsilon_{j'} = +1$.

On the other hand, if $\varepsilon_j \varepsilon_{j'} = -1$, then the argument becomes more involved. But we shall be brief. We may assume that $\varepsilon_j = -1$ and

$\varepsilon_{j'} = +1$. We use again the function ψ_j^- defined by (3.2.9). To compensate for the automorphic factor in (3.2.10) we introduce the Eisenstein series of weight -2:

$$E_{-1}(z,s) = \sum_{\gamma \in \Gamma_\infty \backslash \Gamma} (\operatorname{Im}\gamma(z))^s (j(\gamma,z)/|j(\gamma,z)|)^2.$$

Then we put

$$V_{j,j'}^-(s) = \int_{\mathcal{F}} \psi_j^-(z)\overline{\psi_{j'}(z)} E_{-1}(z,s) d\mu(z).$$

The unfolding method gives

$$\begin{aligned} V_{j,j'}^-(s) &= \int_0^\infty \int_{-\frac{1}{2}}^{\frac{1}{2}} \psi_j^-(z)\overline{\psi_{j'}(z)} y^{s-2} dx dy \\ &= 2i(2\pi)^{-s} R_{j,j'}(s) \int_0^\infty K_{i\kappa}(y) K_{i\kappa'}(y) y^s dy, \end{aligned}$$

since the constant term in the Fourier expansion of $(\partial\psi_j/\partial y)\overline{\psi_{j'}}$ vanishes. Thus we have, for $\operatorname{Re} s > 1$,

$$V_{j,j'}^-(s) = 2i(2\pi)^{-s}\Gamma(s+1; i\kappa, i\kappa') R_{j,j'}(s). \tag{3.2.19}$$

As to $E_{-1}(z,s)$, its Fourier expansion is obtained in just the same way as for $E(z,s)$ (see Section 1.1), and we have

$$\begin{aligned} E_{-1}(z,s) &= y^s + (1 - 1/s)\varphi_\Gamma(s) y^{1-s} \\ &\quad - \frac{\pi^{s+1} y^2}{\Gamma(s+1)\zeta(2s)} \sum_{n \neq 0} |n|^{s+1} \sigma_{1-2s}(|n|) W_s(2\pi|n|y) e(nx), \end{aligned}$$

where φ_Γ is as in (1.1.10), and

$$W_s(y) = (y/2)^{-s-1}\Gamma(s+1) \int_{-\infty}^\infty (1 - i\xi)^2 (1 + \xi^2)^{-s-1} e^{iy\xi} d\xi \quad (\operatorname{Re} s > \tfrac{1}{2}).$$

We have

$$W_s(y) = \int_{-\infty}^\infty (1 - i\xi)^2 e^{iy\xi} \int_0^\infty \eta^s \exp(-\tfrac{1}{2} y(1 + \xi^2)\eta) d\eta d\xi,$$

which converges absolutely. Thus we have

$$W_s(y) = \int_0^\infty \eta^s \exp(-\tfrac{1}{2} y(\eta + \eta^{-1})) \int_{-\infty}^\infty (1 + \eta^{-1} - i\xi)^2 \exp(-\tfrac{1}{2} y\eta\xi^2) d\xi d\eta.$$

We observe that

$$W_s(y) = W_{1-s}(y) \quad (y > 0),$$

which can be confirmed by performing the transformation $\eta \mapsto \eta^{-1}$ and

$\xi \mapsto \eta\xi$. Also it is obvious that $W_s(y)$ is entire in s, and of exponential decay as y increases. Hence $E_{-1}(z;s)$ is regular for $\mathrm{Re}\, s \geq \frac{1}{2}$, and satisfies the functional equation

$$\pi^{-s}\Gamma(s+1)\zeta(2s)E_{-1}(z,s) = \pi^{s-1}\Gamma(2-s)\zeta(2(1-s))E_{-1}(z,1-s).$$

These give the assertions, save for (3.2.13), of the lemma when $\varepsilon_j\varepsilon_{j'} = -1$.

It remains to prove the estimate (3.2.13): By (3.1.18) we have $R_{j,j'}(s) \ll |\rho_j(1)\rho_{j'}(1)|$ for $\mathrm{Re}\, s \geq 2$. If $\varepsilon_j\varepsilon_{j'} = +1$ then the combination of (1.1.9), (3.2.14), and (3.2.18) yields that $s(s-1)R^*_{j,j'}(s)$ is entire. Hence the convexity argument and (3.2.11) give the bound immediately. The case $\varepsilon_j\varepsilon_{j'} = -1$ can be treated in much the same way; and we end the proof of Lemma 3.5.

We shall need also the counterpart of the above assertions for holomorphic cusp-forms: The Hecke L-function attached to $\psi_{j,k}$ (see (3.1.20)) is defined by

$$H_{j,k}(s) = \sum_{n=1}^{\infty} t_{j,k}(n)n^{-s}. \tag{3.2.20}$$

We have

Lemma 3.6 *The function $H_{j,k}(s)$ is entire, and satisfies the functional equation*

$$H_{j,k}(s) = (-1)^k\pi^{2s-1}\frac{\Gamma(k-s+\frac{1}{2})}{\Gamma(k+s-\frac{1}{2})}H_{j,k}(1-s). \tag{3.2.21}$$

In particular we have

$$H_{j,k}(s) \ll k^c \tag{3.2.22}$$

uniformly for bounded s, where c depends only on $\mathrm{Re}\, s$. *Also we have, in the region of absolute convergence,*

$$H_{j,k}(s) = \prod_p \left(1 - t_{j,k}(p)p^{-s} + p^{-2s}\right)^{-1} \tag{3.2.23}$$

as well as

$$H_{j,k}(s)H_{j,k}(s-\alpha)/\zeta(2s-\alpha) = \sum_{n=1}^{\infty} \sigma_\alpha(n)t_{j,k}(n)n^{-s}. \tag{3.2.24}$$

Proof The second assertion follows from (3.1.22), (3.2.21), via the convexity principle. The last two identities can be proved as before. The functional equation follows from the representation

$$\begin{aligned}
&\rho_{j,k}(1)(2\pi)^{\frac{1}{2}-k-s}\Gamma(s+k-\tfrac{1}{2})H_{j,k}(s)\\
&=\int_0^\infty \psi_{j,k}(iy)y^{k+s-\frac{3}{2}}dy\\
&=\int_1^\infty \psi_{j,k}(iy)\{y^{k+s-\frac{3}{2}}+(-1)^k y^{k-s-\frac{1}{2}}\}dy,
\end{aligned}$$

where the last line is due to the relation $\psi_{j,k}(i/y) = (-1)^k y^{2k}\psi_{j,k}(iy)$, $y > 0$. This ends the proof of the lemma.

3.3 Bilinear forms of L-functions

In the subsequent sections we shall investigate the spectro-statistical aspect of the value distribution of Hecke L-functions. We shall, however, restrict ourselves to the simplest situation that is appropriate for our later applications. We shall retain the notations and conventions introduced in the previous section. In addition we put

$$\alpha_j = |\rho_j(1)|^2/\cosh(\pi\kappa_j). \tag{3.3.1}$$

We note that (2.3.2) implies that

$$\sum_{\kappa_j\le K}\alpha_j \ll K^2. \tag{3.3.2}$$

Our aim is to study the sum

$$\mathscr{H}(f;h)=\sum_{j=1}^\infty \alpha_j H_j(\tfrac{1}{2})^2 t_j(f)h(\kappa_j) \quad (f\ge 1), \tag{3.3.3}$$

where the function h is to satisfy some reasonable conditions. More generally we may consider

$$\sum_{j=1}^\infty \alpha_j H_j(u)H_j(v)t_j(f)h(\kappa_j) \quad (f\ge 1), \tag{3.3.4}$$

with arbitrary complex parameters u, v. But, exploiting the special situation in our problem (3.3.3) that

$$H_j(\tfrac{1}{2})=0 \quad \text{if} \quad \varepsilon_j=-1, \tag{3.3.5}$$

which follows from the functional equation (3.2.4), we consider, instead,

$$H(u,v;f;h)=\sum_{j=1}^\infty \varepsilon_j\alpha_j H_j(u)H_j(v)t_j(f)h(\kappa_j) \quad (f\ge 1). \tag{3.3.6}$$

Obviously we have

$$\mathscr{H}(f;h) = H(\tfrac{1}{2}, \tfrac{1}{2}; f; h). \tag{3.3.7}$$

The advantage of $H(u,v;f;h)$ over the sum (3.3.4) will soon be explained. Here we remark only that we shall later prove a theorem of the non-vanishing type and for that particular purpose we shall need a good asymptotic expression for $\mathscr{H}(f;h)$; and the introduction of the parity factor ε_j in (3.3.6) facilitates the relevant analysis considerably.

As to the weight h, we shall assume throughout this section that it is an even entire function such that

$$h(\pm\tfrac{1}{2}i) = 0 \tag{3.3.8}$$

and

$$h(r) \ll \exp(-c|r|^2) \quad (c > 0) \tag{3.3.9}$$

in any fixed horizontal strip. Although these appear quite drastic, they are sufficient for the applications we have in mind. We note that $H(u,v;f;h)$ is obviously entire over $\mathbb{C}^2$ (see (3.2.5)).

Now the identity (3.2.7) gives, in the region of absolute convergence,

$$H(u,v;f;h) = \zeta(u+v)H^{(1)}(u,v;f;h), \tag{3.3.10}$$

where

$$H^{(1)}(u,v;f;h) = \sum_{m=1}^{\infty} m^{-u}\sigma_{u-v}(m) \sum_{j=1}^{\infty} \varepsilon_j \alpha_j t_j(f) t_j(m) h(\kappa_j).$$

We then apply the trace formula (2.5.6) to the inner sum, getting

$$\begin{aligned} &H^{(1)}(u,v;f;g) \\ &\quad = H^{(2)}(u,v,f;g) - \frac{1}{\pi\zeta(u+v)} \int_{-\infty}^{\infty} \frac{\sigma_{2ir}(f)}{f^{ir}|\zeta(1+2ir)|^2} \\ &\quad \times \zeta(u+ir)\zeta(v+ir)\zeta(u-ir)\zeta(v-ir)h(r)dr, \end{aligned} \tag{3.3.11}$$

where

$$H^{(2)}(u,v;f;h) = \sum_{m=1}^{\infty} m^{-u}\sigma_{u-v}(m) \sum_{l=1}^{\infty} \frac{1}{l} S(m,-f;l)\psi\left(\frac{4\pi}{l}\sqrt{mf}\right) \tag{3.3.12}$$

with

$$\psi(x) = 4\pi^{-2} \int_{-\infty}^{\infty} r \sinh(\pi r) K_{2ir}(x) h(r) dr.$$

To transform $H^{(2)}$ we shall consider the function ψ first. On noting that (2.5.2) gives, for $\alpha > 0$,

$$K_{2ir}(x) = \frac{1}{8\sinh(\pi r)} \int_{(\alpha)} \left\{ \frac{\Gamma(s+ir)}{\Gamma(1-s+ir)} - \frac{\Gamma(s-ir)}{\Gamma(1-s-ir)} \right\} \frac{(x/2)^{-2s}}{\cos(\pi s)} ds$$

and that $h(r)$ is even, we have

$$\psi(x) = \pi^{-2} \int_{(\alpha)} \frac{\hat{h}(s)}{\cos(\pi s)} \left(\frac{x}{2}\right)^{-2s} ds \quad (0 < \alpha < \tfrac{1}{2}),$$

where

$$\hat{h}(s) = \int_{-\infty}^{\infty} rh(r) \frac{\Gamma(s+ir)}{\Gamma(1-s+ir)} dr. \tag{3.3.13}$$

We move the path in the last integral to $\operatorname{Im} r = -C$ with an arbitrary $C > 0$, and get the analytic continuation

$$\hat{h}(s) = \int_{\operatorname{Im} r = -C} rh(r) \frac{\Gamma(s+ir)}{\Gamma(1-s+ir)} dr \quad (\operatorname{Re} s > -C). \tag{3.3.14}$$

In particular $\hat{h}$ is entire. We then observe that

$$\hat{h}(\pm\tfrac{1}{2}) = 0, \tag{3.3.15}$$

which will become significant in our later discussion. The case $s = \frac{1}{2}$ is an easy consequence of the definition (3.3.13). On the other hand (3.3.14) implies that

$$\hat{h}(-\tfrac{1}{2}) = -\int_{\operatorname{Im} r = -1} rh(r)(r^2 + \tfrac{1}{4})^{-1} dr.$$

We shift the path back to the real axis while noting the assumption (3.3.8); and get the assertion (3.3.15). It should be remarked that (3.3.14) implies that $\hat{h}(s)$ is of polynomial order if $\operatorname{Re} s$ is bounded. This fact will be used in the sequel without mentioning it explicitly.

Also we may now explain the effect of the presence of the parity symbol ε_j in (3.3.6). This is to gain the exponential decay of the integrand in the definition of ψ, as is clearly visible there. If we had begun discussion with (3.3.4) then we would have had to appeal to the other trace formula (2.3.16), which produces a similar formula to (3.3.11) for the sum (3.3.4), but the relevant transformation of h corresponding to ψ is difficult to handle because its integrand is not of rapid decay.

Now by (3.3.15) we have

$$\psi(x) = \pi^{-2} \int_{(\alpha)} \frac{\hat{h}(s)}{\cos(\pi s)} \left(\frac{x}{2}\right)^{-2s} ds \quad (-\tfrac{3}{2} < \alpha < \tfrac{3}{2}).$$

This implies that we have, in (3.3.12),

$$\sum_{l=1}^{\infty} \frac{1}{l} S(m,-f;l)\psi\left(\frac{4\pi}{l}\sqrt{mf}\right)$$

$$= \pi^{-2} \sum_{l=1}^{\infty} \frac{1}{l} S(m,-f;l) \int_{(\alpha)} \frac{\hat{h}(s)}{\cos(\pi s)} \left(\frac{2\pi}{l}\sqrt{mf}\right)^{-2s} ds, \qquad (3.3.16)$$

where

$$-\tfrac{3}{2} < \alpha < -\tfrac{1}{4}. \qquad (3.3.17)$$

The right side of (3.3.16) converges absolutely because of (2.1.4). We then assume that

$$\mathrm{Re}\, u,\ \mathrm{Re}\, v > 1 - \alpha. \qquad (3.3.18)$$

On this assumption we insert (3.3.16) into (3.3.12), and get

$$H^{(2)}(u,v;f;h) = \pi^{-2} \sum_{l=1}^{\infty} l^{-1} P(u,v;f,l), \qquad (3.3.19)$$

where

$$P(u,v;f,l) = \int_{(\alpha)} \left(\frac{2\pi}{l}\sqrt{f}\right)^{-2s} \frac{\hat{h}(s)}{\cos(\pi s)} Q(s;u,v;f,l)ds \qquad (3.3.20)$$

with

$$Q(s;u,v;f,l) = \sum_{m=1}^{\infty} m^{-u-v} S(m,-f;l)\sigma_{u-v}(m).$$

The expression (3.3.19) holds throughout the region (3.3.18) with (3.3.17), for the right side converges absolutely.

On the other hand, we have

$$Q(s;u,v;f,l) = \sum_{\substack{a=1 \\ (a,l)=1}}^{l} e(-fa/l)D(u+s, u-v; e(a^*/l)), \qquad (3.3.21)$$

where $aa^* \equiv 1 \bmod l$. The D-function is defined by

$$D(s,\xi;e(h/l)) = \sum_{n=1}^{\infty} \sigma_\xi(n)e(nh/l)n^{-s} \quad ((h,l)=1). \qquad (3.3.22)$$

We are going to shift the path in (3.3.20). For this purpose we require some analytical properties of the D-function, which are given in

Lemma 3.7 *The function $D(s,\xi;e(h/l))$ of two complex variables s and ξ is meromorphic over the whole of $\mathbb{C}^2$. As a function of the single variable s it has two simple poles at $s=1$ and $1+\xi$ with residues $l^{\xi-1}\zeta(1-\xi)$ and $l^{-\xi-1}\zeta(1+\xi)$, respectively, provided $\xi\neq 0$; and it is regular elsewhere. Also it satisfies the functional equation*

$$\begin{aligned} D(s,\xi;e(h/l)) &= 2(2\pi)^{2s-\xi-2}l^{\xi-2s+1}\Gamma(1-s)\Gamma(1+\xi-s)\\ &\quad\times\{D(1-s,-\xi;e(h^*/l))\cos(\tfrac{1}{2}\pi\xi)\\ &\quad-D(1-s,-\xi;e(-h^*/l))\cos(\pi(s-\tfrac{1}{2}\xi))\}, \end{aligned} \tag{3.3.23}$$

where $hh^*\equiv 1 \bmod l$. Further, we have

$$D(s,\xi;e(h/l)) \ll \begin{cases} (\log|s|)^2 & \text{if } a(\xi)<\sigma,\\ (l|s|)^{a(\xi)-\sigma}(\log|s|)^2 & \text{if } b(\xi)\le\sigma\le a(\xi),\\ (l|s|)^{a(\xi)-2\sigma}(\log|s|)^2 & \text{if } \sigma<b(\xi), \end{cases} \tag{3.3.24}$$

where $\operatorname{Re}s=\sigma$ and

$$a(\xi)=1+\tfrac{1}{2}(|\operatorname{Re}\xi|+\operatorname{Re}\xi),\quad b(\xi)=-\tfrac{1}{2}(|\operatorname{Re}\xi|-\operatorname{Re}\xi),$$

provided s tends to infinity in an arbitrary fixed vertical strip while ξ remains bounded.

Proof The last assertion is a consequence of the functional equation and the Phragmén–Lindelöf convexity principle. Other assertions follow from the identity

$$D(s,\xi;e(h/l)) = l^{\xi-2s}\sum_{a,b=1}^{l} e(abh/l)\zeta(s,a/l)\zeta(s-\xi,b/l), \tag{3.3.25}$$

where

$$\zeta(s,a)=\sum_{\substack{n=-\infty\\ n+a>0}}^{\infty}(n+a)^{-s}\quad(\operatorname{Re}s>1;\ a\in\mathbb{R})$$

is the Hurwitz zeta-function. As is well-known, $\zeta(s,a)$ is a meromorphic function which is regular for all s save for the simple pole at $s=1$ with residue 1, and satisfies the relation

$$\zeta(s,a) = 2^s\pi^{s-1}\Gamma(1-s)\sum_{n=1}^{\infty} n^{s-1}\sin(\tfrac{1}{2}\pi s + 2n\pi a) \quad (\mathrm{Re}\, s < 0). \tag{3.3.26}$$

Inserting these facts into (3.3.25) we end the proof.

Thus $Q(s;u,v;f,l)$ is a meromorphic function of all s, u, v. We introduce the following sub-region of (3.3.18):

$$-\beta > \mathrm{Re}\, u, \mathrm{Re}\, v > 1-\alpha, \tag{3.3.27}$$

where α, β are such that

$$-\tfrac{3}{2} < \beta < \alpha - 1, \quad -\tfrac{1}{2} < \alpha < -\tfrac{1}{4}. \tag{3.3.28}$$

Then we move the path in (3.3.20) to $\mathrm{Re}\, s = \beta$. According to Lemma 3.7 we see that provided $u \neq v$ the function $Q(s;u,v;f,l)$ has simple poles at $s = 1-u$ and $s = 1-v$ with residues $c_l(f)\zeta(1-u+v)l^{u-v-1}$ and $c_l(f)\zeta(1-v+u)l^{v-u-1}$, respectively, where $c_l(f)$ is the Ramanujan sum defined by (1.1.15). Hence we have, given $u \neq v$,

$$\begin{aligned} &P(u,v;f,l) \\ &= -2\pi i c_l(f) l^{1-u-v}\{(2\pi\sqrt{f})^{2(u-1)}\hat{h}(1-u)\zeta(1-u+v)/\cos(\pi u) \\ &+ (2\pi\sqrt{f})^{2(v-1)}\hat{h}(1-v)\zeta(1-v+u)/\cos(\pi v)\} \\ &+ P^{(1)}(u,v;f,l), \end{aligned} \tag{3.3.29}$$

where $P^{(1)}$ has the same expression as (3.3.20) but with the path $\mathrm{Re}\, s = \beta$. Since we have now $\mathrm{Re}\, u + \beta < 0$, $\mathrm{Re}\, v + \beta < 0$ by (3.3.27) we may replace the factor Q in the integrand of $P^{(1)}$ by the absolutely convergent Dirichlet series obtainable from the functional equation (3.3.23) and the relation (3.3.21). After a rearrangement, we get

$$\begin{aligned} P^{(1)}(u,v;f,l) &= 2(2\pi)^{u+v-2} l^{-u-v+1} \\ &\times \Big\{\sum_{m=1}^{\infty} m^{u-1}\sigma_{v-u}(m) c_l(m+f)\Psi_+(u,v;m/f;h) \\ &+ \sum_{m=1}^{\infty} m^{u-1}\sigma_{v-u}(m) c_l(m-f)\Psi_-(u,v;m/f;h)\Big\}, \end{aligned} \tag{3.3.30}$$

where

$$\begin{aligned} \Psi_+(u,v;x;h) = &-\int_{(\beta)} \Gamma(1-u-s)\Gamma(1-v-s) \\ &\times \cos(\pi(s+\tfrac{1}{2}(u+v)))\frac{\hat{h}(s)}{\cos(\pi s)} x^s ds \end{aligned} \tag{3.3.31}$$

and

$$\Psi_-(u,v;x;h) = \cos(\tfrac{1}{2}\pi(u-v)) \int_{(\beta)} \Gamma(1-u-s)\Gamma(1-v-s)\frac{\hat{h}(s)}{\cos(\pi s)}x^s ds. \tag{3.3.32}$$

While retaining the condition (3.3.27) with (3.3.28) we insert (3.3.30) into (3.3.29) and the result into (3.3.19); because of absolute convergence we may exchange the order of sums freely. Then, invoking (1.1.14), we obtain, via (3.3.10)–(3.3.11), the following transformation of the sum (3.3.6):

$$\begin{aligned}
&H(u,v;f;h)\\
&= -\pi^{-1}\int_{-\infty}^{\infty} \frac{\sigma_{2ir}(f)\zeta(u+ir)\zeta(u-ir)\zeta(v+ir)\zeta(v-ir)}{f^{ir}|\zeta(1+2ir)|^2}h(r)dr\\
&\quad + 2(\pi i)^{-1}(2\pi\sqrt{f})^{2(u-1)}\frac{\hat{h}(1-u)}{\cos(\pi u)}\sigma_{1-u-v}(f)\zeta(1-u+v)\\
&\quad + 2(\pi i)^{-1}(2\pi\sqrt{f})^{2(v-1)}\frac{\hat{h}(1-v)}{\cos(\pi v)}\sigma_{1-u-v}(f)\zeta(1-v+u)\\
&\quad + 8(2\pi)^{u+v-4}\sum_{m=1}^{\infty} m^{u-1}\sigma_{v-u}(m)\sigma_{1-u-v}(m+f)\Psi_+(u,v;m/f;h)\\
&\quad + 8(2\pi)^{u+v-4}\sum_{\substack{m=1\\ m\neq f}}^{\infty} m^{u-1}\sigma_{v-u}(m)\sigma_{1-u-v}(m-f)\Psi_-(u,v;m/f;h)\\
&\quad + 8(2\pi)^{u+v-4}f^{u-1}\sigma_{v-u}(f)\zeta(u+v-1)\Psi_-(u,v;1;h).
\end{aligned}$$

We have proved this given the conditions $u \neq v$ and (3.3.27) with (3.3.28). But the first condition can be dropped by an obvious convention; and the second can be relaxed to $1+\beta < \operatorname{Re} u, \operatorname{Re} v < -\beta$ with $-\frac{3}{2} < \beta < -\frac{1}{2}$ where β is involved in the definition of $\Psi_\pm$, provided that the integrated term on the right side of the last identity is replaced by an appropriate analytic continuation. In fact both sums over m in the last expression converge absolutely in this new domain of u, v; so, in particular, we may set $(u,v) = (\frac{1}{2}, \frac{1}{2})$ in those sums without any modification. Further, the integrated term has the following continuation to the domain $\{\operatorname{Re} u, \operatorname{Re} v < 1\}$:

$$\begin{aligned}
&-\pi^{-1}\int_{-\infty}^{\infty} \frac{\sigma_{2ir}(f)\zeta(u+ir)\zeta(u-ir)\zeta(v+ir)\zeta(v-ir)}{f^{ir}|\zeta(1+2ir)|^2}h(r)dr\\
&- 4\sigma_{2(u-1)}(f)f^{1-u}\zeta(u+v-1)\zeta(v-u+1)h(i(u-1))/\zeta(3-2u)\\
&- 4\sigma_{2(v-1)}(f)f^{1-v}\zeta(u+v-1)\zeta(u-v+1)h(i(v-1))/\zeta(3-2v).
\end{aligned}$$

To show this we move the path in the original integral to $\operatorname{Im} r = C$ with a sufficiently large non-integral $C > 0$, passing over the poles $r = i(u-1)$, $i(v-1)$ and those corresponding to the zeros of $\zeta(1+2ir)$. This gives an analytic continuation to the domain containing $\{1-C < \operatorname{Re} u, \operatorname{Re} v < 1\}$. Then, restricting (u, v) to the latter domain, we shift the new path back to the real axis, passing over the poles at $r = i(1-u)$, $i(1-v)$ and those corresponding to the zeros of $\zeta(1+2ir)$. Computing the relevant residues we get the above continuation. We then observe that (3.3.15) implies

$$\begin{aligned}\lim_{(u,v)\to(\frac{1}{2},\frac{1}{2})} &\Big\{(2\pi\sqrt{f})^{2(u-1)}\frac{\hat{h}(1-u)}{\cos(\pi u)}\sigma_{1-u-v}(f)\zeta(1-u+v) \\ &+ (2\pi\sqrt{f})^{2(v-1)}\frac{\hat{h}(1-v)}{\cos(\pi v)}\sigma_{1-u-v}(f)\zeta(1-v+u)\Big\} \\ &= \pi^{-2}\Big\{(c_E - \log(2\pi\sqrt{f}))(\hat{h})'(\tfrac{1}{2}) + \tfrac{1}{4}(\hat{h})''(\tfrac{1}{2})\Big\}d(f)f^{-\frac{1}{2}}\end{aligned}$$

with the Euler constant c_E, and that (3.3.8) implies

$$\begin{aligned}\lim_{(u,v)\to(\frac{1}{2},\frac{1}{2})} &\{\sigma_{2(u-1)}(f)f^{1-u}\zeta(u+v-1)\zeta(v-u+1)h(i(u-1))/\zeta(3-2u) \\ &+ \sigma_{2(v-1)}(f)f^{1-v}\zeta(u+v-1)\zeta(u-v+1)h(i(v-1))/\zeta(3-2v)\} \\ &= -3i\pi^{-2}h'(-\tfrac{1}{2}i)\sigma_{-1}(f)f^{\frac{1}{2}}.\end{aligned}$$

Combining these, we obtain

Lemma 3.8 *Let $h(r)$ be an even entire function satisfying* (3.3.8) *and* (3.3.9). *Then we have*

$$\mathscr{H}(f;h) = \sum_{\nu=1}^{7}\mathscr{H}_\nu(f;h). \tag{3.3.33}$$

Here

$$\mathscr{H}_1(f;h) = -2\pi^{-3}i\Big\{(c_E - \log(2\pi\sqrt{f}))(\hat{h})'(\tfrac{1}{2}) + \tfrac{1}{4}(\hat{h})''(\tfrac{1}{2})\Big\}d(f)f^{-\frac{1}{2}},$$

$$\mathscr{H}_2(f;h) = \pi^{-3}\sum_{m=1}^{\infty} m^{-\frac{1}{2}}d(m)d(m+f)\Psi^+(m/f;h),$$

$$\mathscr{H}_3(f;h) = \pi^{-3}\sum_{m=1}^{\infty}(m+f)^{-\frac{1}{2}}d(m)d(m+f)\Psi^-(1+m/f;h),$$

$$\mathscr{H}_4(f;h) = \pi^{-3}\sum_{m=1}^{f-1} m^{-\frac{1}{2}}d(m)d(f-m)\Psi^-(m/f;h),$$

$$\mathscr{H}_5(f;h) = -(2\pi^3)^{-1} f^{-\frac{1}{2}} d(f) \Psi^-(1;h),$$

$$\mathscr{H}_6(f;h) = -12\pi^{-2} i \sigma_{-1}(f) f^{\frac{1}{2}} h'(-\tfrac{1}{2}i),$$

$$\mathscr{H}_7(f;h) = -\pi^{-1} \int_{-\infty}^{\infty} \frac{|\zeta(\frac{1}{2}+ir)|^4}{|\zeta(1+2ir)|^2} \sigma_{2ir}(f) f^{-ir} h(r) dr,$$

where $\hat{h}$ *is defined by* (3.3.13), *and*

$$\Psi^+(x;h) = \int_{(\beta)} \Gamma(\tfrac{1}{2}-s)^2 \tan(\pi s) \hat{h}(s) x^s ds, \tag{3.3.34}$$

$$\Psi^-(x;h) = \int_{(\beta)} \Gamma(\tfrac{1}{2}-s)^2 \frac{\hat{h}(s)}{\cos(\pi s)} x^s ds \tag{3.3.35}$$

with $-\frac{3}{2} < \beta < \frac{1}{2}$.

In the next section we shall show various applications of this lemma; and for that purpose we have to know a little more about $(\hat{h})'(\frac{1}{2})$, $(\hat{h})''(\frac{1}{2})$, $\Psi^{\pm}(x;h)$: The definition (3.3.13) gives, for $\mathrm{Re}\, s > 0$,

$$(\hat{h})'(s) = \int_{-\infty}^{\infty} rh(r) \frac{\Gamma(s+ir)}{\Gamma(1-s+ir)} \left\{ \frac{\Gamma'}{\Gamma}(s+ir) + \frac{\Gamma'}{\Gamma}(1-s+ir) \right\} dr. \tag{3.3.36}$$

Thus we have

$$(\hat{h})'(\tfrac{1}{2}) = 2 \int_{-\infty}^{\infty} rh(r) \frac{\Gamma'}{\Gamma}(\tfrac{1}{2}+ir) dr; \tag{3.3.37}$$

and differentiating (3.3.36) we have

$$(\hat{h})''(\tfrac{1}{2}) = 4 \int_{-\infty}^{\infty} rh(r) \left\{ \frac{\Gamma'}{\Gamma}(\tfrac{1}{2}+ir) \right\}^2 dr. \tag{3.3.38}$$

As to Ψ^+, we shift the path in (3.3.34) to $\mathrm{Re}\, s = \frac{1}{4}$, and insert (3.3.13) into the result. Exchanging the order of integration we get

$$\Psi^+(x;h) = (\pi i)^{-1} \int_{-\infty}^{\infty} rh(r) \sinh(\pi r) \times \int_{(\frac{1}{4})} \Gamma(\tfrac{1}{2}-s)^2 \Gamma(s+ir) \Gamma(s-ir) \sin(\pi s) x^s ds dr. \tag{3.3.39}$$

To transform this we assume first that $x > 1$. Shifting the path to

$\operatorname{Re} s = -\infty$ in the inner integral we see that it is equal to

$$\frac{\pi^2 i}{\cosh(\pi r)}\left\{\frac{\Gamma(\frac{1}{2}+ir)^2}{\Gamma(1+2ir)}F(\tfrac{1}{2}+ir,\tfrac{1}{2}+ir;1+2ir;-1/x)x^{-ir}\right. \\ \left.+\frac{\Gamma(\frac{1}{2}-ir)^2}{\Gamma(1-2ir)}F(\tfrac{1}{2}-ir,\tfrac{1}{2}-ir;1-2ir;-1/x)x^{ir}\right\}$$

with the hypergeometric function F. Here we invoke an integral representation of the hypergeometric function: We have, for $|y|<1$,

$$F(a,b;c;y) = \frac{\Gamma(c)}{\Gamma(a)\Gamma(c-a)}\int_0^1 x^{a-1}(1-x)^{c-a-1}(1-xy)^{-b}dx, \quad (3.3.40)$$

provided $\operatorname{Re} c > \operatorname{Re} a > 0$. Thus we have

$$\Psi^+(x;h) = 2\pi\int_0^1 \{y(1-y)(1+y/x)\}^{-\frac{1}{2}} \\ \times\int_{-\infty}^{\infty} rh(r)\tanh(\pi r)\left\{\frac{y(1-y)}{x+y}\right\}^{ir} drdy. \quad (3.3.41)$$

As the expression (3.3.39) shows clearly, $\Psi^+(x;h)$ is regular for $\operatorname{Re} x > 0$; and the right side of the last formula is also regular there. Hence by analytic continuation (3.3.41) holds, in particular, for $x > 0$. Moving to Ψ^-, we have, by the definitions (3.3.13) and (3.3.35),

$$\Psi^-(x;h) = (\pi i)^{-1}\int_{-\infty}^{\infty} rh(r)\sinh(\pi r) \\ \times\int_{(\frac{1}{4})} \Gamma(s+ir)\Gamma(s-ir)\Gamma(\tfrac{1}{2}-s)^2 x^s dsdr. \quad (3.3.42)$$

We consider three cases separately according to whether x is greater than, equal to, or less than 1. Thus, when $x > 1$, we argue in just the same way as above and have

$$\Psi^-(x;h) = 2\pi i\int_0^1 \{y(1-y)(1-y/x)\}^{-\frac{1}{2}} \\ \times\int_{-\infty}^{\infty} \frac{rh(r)}{\cosh(\pi r)}\left\{\frac{y(1-y)}{x-y}\right\}^{ir} drdy. \quad (3.3.43)$$

When $x = 1$, we appeal to (2.6.13), which gives immediately

$$\Psi^-(1;h) = 2\pi^2\int_{-\infty}^{\infty} rh(r)\frac{\sinh(\pi r)}{(\cosh(\pi r))^2}dr. \quad (3.3.44)$$

Finally, when $0 < x < 1$, we do not use (3.3.42) but transform (3.3.35) in a different way. By (3.3.14) we have, for $-\frac{3}{2} < \beta < \frac{1}{2}$,

$$\begin{aligned}\Psi^-(x;h) &= \int_{(\beta)} x^s \frac{\Gamma(\frac{1}{2}-s)^2}{\cos(\pi s)} \int_{\mathrm{Im}\, r=-2} rh(r) \frac{\Gamma(s+ir)}{\Gamma(1-s+ir)} drds \\ &= \int_{(\beta)} x^s \frac{\Gamma(\frac{1}{2}-s)^2}{\cos(\pi s)} \int_{\mathrm{Im}\, r=-2} rh(r)\Gamma(1-2s)^{-1} \\ &\times \int_0^\infty y^{s+ir-1}(1+y)^{s-ir-1} dydrds.\end{aligned}$$

This triple integral converges absolutely, provided $-\frac{3}{2} < \beta < \frac{1}{2}$, $\beta \neq -\frac{1}{2}$. Hence we have, for such a β,

$$\begin{aligned}\Psi^-(x;h) = \int_0^\infty &\left\{ \int_{(\beta)} x^s (y(y+1))^{s-1} \frac{\Gamma(\frac{1}{2}-s)^2}{\Gamma(1-2s)\cos(\pi s)} ds \right\} \\ &\times \left\{ \int_{-\infty}^{\infty} rh(r) \left(\frac{y}{1+y}\right)^{ir} dr \right\} dy. \qquad (3.3.45)\end{aligned}$$

3.4 Spectral mean values

As the first application of Lemma 3.8 we shall consider the estimation of the expression

$$\mathscr{B}(K,G) = \sum_{K \le \kappa_j < K+G} \alpha_j H_j(\tfrac{1}{2})^2 \left| \sum_{p=1}^{P} b_p t_p^{i\kappa_j} \right|^2.$$

Sums of this type will appear in our discussion to be developed in the final chapter. Here $\{b_p\}$ is an arbitrary complex vector, and we assume that

$$C(\log K)^{\frac{1}{2}} < G < K(\log K)^{-1}, \qquad (3.4.1)$$

$$0 < T \le t_1 < t_2 < \cdots < t_P \le 2T, \quad \log T \approx \log K, \qquad (3.4.2)$$

where K is a large parameter, and C is a large positive constant.

Theorem 3.1 *In addition to* (3.4.1) *and* (3.4.2) *let us assume that* $|t_p - t_q| > \eta > 0$ *for* $p \neq q$. *Then we have*

$$\sum_{K \le \kappa_j < K+G} \alpha_j H_j(\tfrac{1}{2})^2 \left| \sum_{p=1}^{P} b_p t_p^{i\kappa_j} \right|^2 \ll (G + T\eta^{-1}) \|b_p\|^2 K \log K, \qquad (3.4.3)$$

where $\|b_p\|$ is the Euclidean norm of $\{b_p\}$, and the implied constant depends only on C. In particular we have

$$\sum_{K\le\kappa_j<K+G} \alpha_j H_j(\tfrac{1}{2})^2 \ll GK\log K. \tag{3.4.4}$$

Proof We put

$$\begin{aligned} h_0(r) &= h_0(r;K,G) \\ &= (r^2+\tfrac{1}{4})\{\exp(-((r-K)/G)^2)+\exp(-((r+K)/G)^2)\}. \end{aligned} \tag{3.4.5}$$

Then we have

$$\mathscr{B}(K,G) \ll K^{-2}\mathscr{B}_1(K,G), \tag{3.4.6}$$

where

$$\mathscr{B}_1(K,G) = \sum_{j=1}^{\infty} \alpha_j H_j(\tfrac{1}{2})^2 h_0(\kappa_j)\Big\{\Big|\sum_{p=1}^{P} b_p t_p^{i\kappa_j}\Big|^2 + \Big|\sum_{p=1}^{P} \overline{b_p} t_p^{i\kappa_j}\Big|^2\Big\}.$$

Expanding out the squares we have

$$\mathscr{B}_1(K,G) = 2\sum_{p,q=1}^{P} b_p\overline{b_q}\mathscr{H}(1;h_0(\cdot;t_p/t_q)),$$

where

$$h_0(r,\xi) = h_0(r)\cos(r\log\xi).$$

Obviously this function satisfies the conditions supposed in Lemma 3.8.

We then proceed to the estimation of $\Psi^{\pm}$ on the assumption (3.4.1). We have, by (3.3.41),

$$\Psi^+(m;h_0(\cdot;\xi)) = \tfrac{1}{2}\{\psi^+(m,\xi)+\psi^+(m,\xi^{-1})\},$$

where

$$\begin{aligned} \psi^+(m,\xi) = 2\pi\int_0^1 &\{y(1-y)(1+y/m)\}^{-\frac{1}{2}} \\ &\times \int_{-\infty}^{\infty} rh_0(r)\tanh(\pi r)\Big\{\frac{\xi y(1-y)}{m+y}\Big\}^{ir} drdy. \end{aligned}$$

In the inner integral we shift the path to $\operatorname{Im} r = -1$. The pole at $r = -\frac{1}{2}i$ of $\tanh(\pi r)$ is canceled by the zero of $h_0(r)$, so the result is

$$\int_{-\infty}^{\infty} (r-i)h_0(r-i)\tanh(\pi r)\Big\{\frac{\xi y(1-y)}{m+y}\Big\}^{ir+1} dr.$$

We divide this into two parts: the one corresponding to $|K \pm r| \geq \frac{1}{2} G \log K$ and the rest. The first part is obviously $O(\xi m^{-1} \exp(-\frac{1}{5}(\log K)^2))$. In the second part we have $\tanh(\pi r) = \operatorname{sgn}(r) + O(e^{-K})$ by (3.4.1); this O-term contributes negligibly. Thus the last integral is equal to

$$2\mathrm{Re} \int_{-\infty}^{\infty} r h_0(r) \left\{ \frac{\xi y(1-y)}{m+y} \right\}^{ir} dr + O\left(\xi m^{-1} \exp(-\tfrac{1}{5}(\log K)^2)\right)$$
$$\ll K^3 G \exp\left(-\tfrac{1}{8}\left(G \log \frac{\xi y(1-y)}{m+y} \right)^2 \right) + O\left(\xi m^{-1} \exp(-\tfrac{1}{5}(\log K)^2)\right).$$

We note that

$$\sup_{0 \leq y \leq 1} \frac{\xi y(1-y)}{m+y} < \frac{\xi}{4m}.$$

Hence we have, on noting the assumptions (3.4.1) and (3.4.2),

$$\begin{aligned} \Psi^{+}(m; h_0(\cdot\,; t_p/t_q)) &\ll K^3 G \exp(-\tfrac{1}{8}(G \log(2m))^2) \\ &\quad + m^{-1} \exp(-\tfrac{1}{5}(\log K)^2) \\ &\ll m^{-1} K^{-C} \end{aligned}$$

uniformly in $m \geq 1$.

As to $\Psi^{-}(m; h_0(\cdot, t_p/t_q))$ we use (3.3.43) when $m > 1$. Shifting the path in the inner integral to $\mathrm{Im}\, r = -1$ we get immediately

$$\Psi^{-}(m; h_0(\cdot, t_p/t_q)) \ll m^{-1} \exp(-(\log K)^2)$$

uniformly for $m \geq 2$. Obviously this holds for $m = 1$, too, for we have (3.3.44).

From these and (3.3.33) we get

$$\mathscr{B}_1(K, G) \leq 2 \sum_{p,q=1}^{P} b_p \overline{b_q} \mathscr{H}_1(1; h_0(\cdot\,; t_p/t_q)) + O(\|b_p\|^2 K^{-C}),$$

where $\|b_p\|$ is as above. In fact, the contribution of the parts corresponding to $\mathscr{H}_\nu$ $(2 \leq \nu \leq 4)$ is absorbed into this error term, and the part $\mathscr{H}_6$ is obviously negligible; moreover the contribution of the continuous spectrum, i.e., the part $\mathscr{H}_7$, is not positive, and can be discarded. On the other hand the last double sum is, by (3.3.37) and (3.3.38),

$$\int_{-\infty}^{\infty} |r| h_0(r) \left| \sum_{p=1}^{P} a_p t_p^{ir} \right|^2 \left\{ \left| \mathrm{Im} \frac{\Gamma'}{\Gamma}(\tfrac{1}{2} + ir) \right| + \left| \mathrm{Im} \left(\frac{\Gamma'}{\Gamma} \right)^2 (\tfrac{1}{2} + ir) \right| \right\} dr$$
$$\ll \int_{-\infty}^{\infty} |r| \log(|r| + 2) h_0(r; K, G) \left| \sum_{p=1}^{P} b_p t_p^{ir} \right|^2 dr,$$

which is

$$\ll GK^3 \log K \sum_{p,q}^{P} |b_p b_q| \exp(-\tfrac{1}{4}(G \log(t_p/t_q))^2)$$

$$\ll GK^3 \log K \sum_{p,q}^{P} |b_p b_q| \{1 + (|t_p - t_q|G/T)^2\}^{-1}.$$

This gives rise to the assertion of the theorem.

Our next aim is to prove a theorem of the non-vanishing type, which will play an important rôle in the final chapter. To this end we shall first give an approximative representation of $H_j(\frac{1}{2})$:

Lemma 3.9 *Let us assume that*

$$|\kappa_j - K| \le G \log K \tag{3.4.7}$$

with

$$(\log K)^2 < G < K^{1-\delta}, \quad \delta > 0. \tag{3.4.8}$$

Then we have, for any $N \ge 1$ and $\lambda = C \log K$ with a sufficiently large $C > 0$,

$$\begin{aligned} H_j(\tfrac{1}{2}) &= \sum_{f \le 3K} t_j(f) f^{-\frac{1}{2}} \exp(-(f/K)^\lambda) \\ &\quad - \sum_{\nu=0}^{N_1} \sum_{f \le 3K} t_j(f) f^{-\frac{1}{2}} U_\nu(fK)(1 - (\kappa_j/K)^2)^\nu \\ &\quad + O(K^{-\frac{1}{3}N} + K^{-\frac{1}{2}C}) \end{aligned} \tag{3.4.9}$$

with the implied constant depending only on δ, C, and N. Here $N_1 = [2N/\delta]$ and

$$U_\nu(x) = \frac{1}{2\pi i \lambda} \int_{(-\lambda^{-1})} (4\pi^2 K^{-2} x)^w u_p(w) \Gamma(w/\lambda) dw, \tag{3.4.10}$$

where $u_p(w)$ is a polynomial of degree $\le 2N_1$, whose coefficients are independent of κ_j and bounded by a constant depending only on δ and N.

Proof We consider the integral

$$\mathscr{R} = \frac{1}{2\pi i \lambda} \int_{(3)} H_j(w + \tfrac{1}{2}) K^w \Gamma(w/\lambda) dw. \tag{3.4.11}$$

We may assume that $\varepsilon_j = 1$ for an obvious reason. We have

$$\mathscr{R} = \sum_{f \le 3K} t_j(f) f^{-\frac{1}{2}} \exp(-(f/K)^\lambda) + O(e^{-K}).$$

On the other hand, shifting the path in (3.4.11) to $\operatorname{Re} w = -\frac{1}{2}\lambda$ and recalling the functional equation (3.2.4) with $\varepsilon_j = 1$, we get

$$\mathscr{R} = H_j(\tfrac{1}{2}) + \sum_{f=1}^{\infty} t_j(f) f^{-\frac{1}{2}} \mathscr{R}_j(fK),$$

where

$$\mathscr{R}_j(x) = \frac{1}{2\pi^2 i\lambda} \int_{(-\frac{1}{2}\lambda)} (4\pi^2 x)^w \Gamma(\tfrac{1}{2} - w + i\kappa_j)\Gamma(\tfrac{1}{2} - w - i\kappa_j)$$
$$\times \left\{ \cosh(\pi\kappa_j) + \sin(\pi w) \right\} \Gamma(w/\lambda) dw. \qquad (3.4.12)$$

By Stirling's formula this integrand is

$$\ll (4\pi^2 x)^{-\frac{1}{2}\lambda} \left(|w + i\kappa_j||w - i\kappa_j|\right)^{\frac{1}{2}\lambda} \exp(-\pi|w|/(2\lambda));$$

and thus

$$\mathscr{R}_j(x) = O((4\pi^2 x K^{-2})^{-\lambda/2}),$$

where the implied constant is absolute. In fact, when $|w| < \sqrt{K}$ we see that the factor $\left(|w + i\kappa_j||w - i\kappa_j|\right)^{\frac{1}{2}\lambda}$ is $O(K^\lambda)$, and when $|w| \ge \sqrt{K}$ the integrand itself is negligible because of the factor $\exp(-\pi|w|/(2\lambda))$ provided (3.4.7) holds. This estimate of $\mathscr{R}_j$ allows us to truncate the last sum over f at $f = [3K]$ with an error of size K^{-C}. In the remaining terms, i.e., those with $f \le K$, we modify $\mathscr{R}_j(fK)$ as follows: We shift the path in (3.4.12) to $\operatorname{Re} w = -\lambda^{-1}$, and restrict the integration to the interval corresponding to $|\operatorname{Im} w| \le \lambda^2$. In the part where $|\operatorname{Im} w| > \lambda^2$ the integrand is clearly $O(\exp(-\pi|w|/(4\lambda)))$; so the error thus caused by this truncation of the integral is in total $O(K^{-\frac{2}{3}C})$. Hence we have, uniformly for all κ_j satisfying (3.4.7),

$$H_j(\tfrac{1}{2}) = \sum_{f \le 3K} t_j(f) f^{-\frac{1}{2}} \exp(-(f/K)^\lambda)$$
$$- \sum_{f \le 3K} t_j(f) f^{-\frac{1}{2}} \mathscr{R}_j^{(1)}(fK) + O(K^{-\frac{1}{2}C}), \qquad (3.4.13)$$

where

$$\mathscr{R}_j^{(1)}(x) = \frac{1}{2\pi^2 i\lambda} \int_{-\lambda^{-1} - i\lambda^2}^{-\lambda^{-1} + i\lambda^2} (4\pi^2 x)^w \Gamma(\tfrac{1}{2} - w + i\kappa_j)\Gamma(\tfrac{1}{2} - w - i\kappa_j)$$
$$\times \left\{ \cosh(\pi\kappa_j) + \sin(\pi w) \right\} \Gamma(w/\lambda) dw. \qquad (3.4.14)$$

Then we use Stirling's formula in a more precise way: We have, for any $N \geq 1$ and for those w relevant to (3.4.14),

$$\log\Gamma(\tfrac{1}{2} - w + i\kappa_j)$$
$$= (-w + i\kappa_j)\log(\tfrac{1}{2} - w + i\kappa_j) - \tfrac{1}{2} + w - i\kappa_j + \tfrac{1}{2}\log(2\pi)$$
$$+ \sum_{\nu=1}^{2N} c_\nu(\tfrac{1}{2} - w + i\kappa_j)^{-\nu} + O(K^{-2N-\frac{1}{2}}),$$

where c_ν's are absolute constants, and the implied constant depends only on N; thus

$$\log\Gamma(\tfrac{1}{2} - w + i\kappa_j) = (-w + i\kappa_j)\{\log(\kappa_j) + \tfrac{1}{2}\pi i\} - i\kappa_j + \tfrac{1}{2}\log(2\pi)$$
$$+ \sum_{\nu=1}^{2N} p_\nu(w)\kappa_j^{-\nu} + O(K^{-2N}(\log K)^{12N+6})$$

with certain polynomials p_ν of degree $\leq \nu + 1$ with constant coefficients. Adding to this the corresponding formula for $\log\Gamma(\frac{1}{2} - w - i\kappa_j)$, we have

$$\log\left\{\Gamma(\tfrac{1}{2} - w + i\kappa_j)\Gamma(\tfrac{1}{2} - w - i\kappa_j)\right\} = -2w\log(\kappa_j) - \pi\kappa_j + \log(2\pi)$$
$$+ \sum_{\nu=1}^{N} p_{2\nu}(w)\kappa_j^{-2\nu} + O(K^{-2N}(\log K)^{12N+6}).$$

This implies readily that the integrand of (3.4.14) can be replaced by

$$\pi(4\pi^2\kappa_j^{-2}x)^w\left\{1 + \sum_{\nu=1}^{N} q_\nu(w)\kappa_j^{-2\nu} + O(K^{-N})\right\}\Gamma(w/\lambda),$$

where $q_\nu(w)$ are polynomials of degree $\leq 3\nu$ with constant coefficients, and the O-constant depends only on N. Then we expand each $\kappa_j^{-2w-2\nu} = K^{-2w-2\nu}(1 - (1 - (\kappa_j/K)^2))^{-w-\nu}$ into a power series in $(1 - (\kappa_j/K)^2) = O(K^{-\delta}\log K)$; and truncate it at the power $N_1 = [2N/\delta]$ by using the following identity: For any complex η and any positive integer P

$$\Gamma(\eta + 1)^{-1}(1 + x)^\eta$$
$$= \sum_{j=0}^{P-1} \frac{x^j}{\Gamma(j + 1)\Gamma(\eta - j + 1)}$$
$$+ \frac{x^P}{\Gamma(P)\Gamma(\eta - P + 1)} \int_0^1 (1 - \theta)^{P-1}(1 + x\theta)^{\eta-P}\, d\theta, \qquad (3.4.15)$$

which is a result of repeated application of integration by parts to the last term.

Rearranging the result of truncation we see that the integrand of (3.4.14) can be written as

$$\pi(4\pi^2K^{-2}x)^w\{Q(w,1-(\kappa_j/K)^2)+O(K^{-N})\}\Gamma(w/\lambda), \qquad (3.4.16)$$

where

$$Q(w,y)=\sum_{\nu=0}^{N_1} u_\nu(w)y^\nu$$

with

$$u_0(w)=1+\sum_{\nu=1}^{N} q_\nu(w)K^{-2\nu}.$$

Inserting (3.4.16) into (3.4.14) and restoring the range of integration to the whole line $\mathrm{Re}\, w=-\lambda^{-1}$, we get, uniformly for $f\le 3K$,

$$\mathscr{R}_j^{(1)}(fK)=\sum_{\nu=0}^{N_1} U_\nu(fK)(1-(\kappa_j/K)^2)^\nu+O(K^{-N}).$$

This and (3.4.13) end the proof of the lemma.

We are now ready to prove

Theorem 3.2 *There are infinitely many κ such that*

$$\sum_{\kappa_j=\kappa}\alpha_j H_j(\tfrac{1}{2})^3>0. \qquad (3.4.17)$$

Proof We shall study more generally the asymptotic behavior of the sum

$$\mathscr{C}(K,G)=\sum_{j=1}^{\infty}\alpha_j H_j(\tfrac{1}{2})^3 h_0(\kappa_j),$$

where $h_0(r)$ is defined by (3.4.5) with (3.4.8). The presence of the factor $h_0(\kappa_j)$ allows us to restrict the range of summation to (3.4.7) with a negligible error. We then apply (3.4.9) to one factor of $H_j(\frac{1}{2})^3$, and subsequently restore the summation to the full range. This gives

$$\begin{aligned}\mathscr{C}(K,G)=&\sum_{f\le 3K} f^{-\frac{1}{2}}\exp(-(f/K)^\lambda)\mathscr{H}(f;h_0)\\ &-\sum_{\nu=0}^{N_1}\sum_{f\le 3K} f^{-\frac{1}{2}}U_\nu(fK)(1-(\kappa_j/K)^2)^\nu\mathscr{H}(f;h_\nu)+O(1),\end{aligned} \qquad (3.4.18)$$

where $\mathscr{H}$ is defined by (3.3.3), and

$$h_\nu(r) = h_0(r)(1-(r/K)^2)^\nu.$$

It is understood here that N is now fixed so that the error estimate in (3.4.18) is attained; and consequently N_1 depends only on δ. We note also that (3.2.5) and (3.3.2) have been used. Since h_ν satisfies the relevant conditions, we may appeal to (3.3.33). Then we have to estimate $\Psi^\pm(m/f;h_\nu)$ for those f appearing in (3.4.18).

We begin with $\Psi^+(m/f;h_\nu)$. As before, we shift the path of the r-integral in (3.3.41) with $h = h_\nu$ to $\mathrm{Im}\, r = -1$ and see that it is equal to

$$\int_{-\infty}^{\infty} (r-i)h_\nu(r-i)\tanh(\pi r)\left\{\frac{y(1-y)}{m/f+y}\right\}^{ir+1} dr.$$

The part corresponding to $|K \pm r| \geq \frac{1}{2}G\log K$ is $O(m^{-1}f\exp(-\frac{1}{4}\times(\log K)^2))$ uniformly for $\nu \leq N_1$. In the remaining part we have $\tanh(\pi r) = \mathrm{sgn}(r) + O(e^{-K})$. Thus the last integral is

$$\ll K^3G(G/K)^\nu \exp\left(-\tfrac{1}{8}\left(\frac{y(1-y)}{m/f+y}\right)^2\right) + m^{-1}f\exp(-\tfrac{1}{4}(\log K)^2). \quad (3.4.19)$$

We note that

$$\sup_{0\leq y\leq 1}\frac{y(1-y)}{m/f+y} = \frac{m/f}{(m/f+((m/f)^2+m/f)^{\frac{1}{2}})^2} < 1.$$

If m/f is small, this is $1-O((m/f)^{\frac{1}{2}})$; and if m/f is large it is $O(f/m)$. Gathering these we get

$$\begin{aligned}\Psi^+(m/f;h_\nu) \ll\ & K^3G(G/K)^\nu\exp(-cG^2m/f)\\ &+ m^{-1}f\exp(-\tfrac{1}{4}(\log K)^2) \quad (m \leq 2f),\end{aligned} \qquad (3.4.20)$$

and

$$\begin{aligned}\Psi^+(m/f;h_\nu) \ll\ & K^3G(G/K)^\nu\exp(-c(G\log(m/f))^2)\\ &+ m^{-1}f\exp(-\tfrac{1}{4}(\log K)^2) \quad (m > 2f),\end{aligned} \qquad (3.4.21)$$

where c is a positive absolute constant.

We next consider $\Psi^-(m/f;h_\nu)$. If $m \geq f$ then we use (3.3.43) and (3.3.44), getting readily

$$\Psi^-(m/f;h_\nu) \ll m^{-1}f\exp(-\tfrac{1}{4}(\log K)^2) \quad (m \geq f). \qquad (3.4.22)$$

If $m < f$ we use (3.3.45). The r-integral with $h = h_\nu$ can be computed explicitly; and we have

$$\Psi^-(m/f;h_\nu) \ll |\cos(\pi\beta)|^{-1}K^3G(G/K)^\nu(m/f)^\beta$$
$$\times \int_0^\infty (y(y+1))^{\beta-1}\exp\left(-\tfrac{1}{8}\left(G\log\frac{y}{1+y}\right)^2\right)dy. \quad (3.4.23)$$

The part of this integral which corresponds to the range $0 < y < G(\log K)^{-1}$ is $O(\exp(-\frac{1}{4}(\log K)^2))$; note that we have used the lower bound of G given in (3.4.8). The integral over the remaining range is $O(G^{2\beta-1})$. Thus, setting $\beta = -\frac{3}{2} + (\log K)^{-1}$, we find that

$$\Psi^-(m/f;h_\nu) \ll K^3G^{-3}(G/K)^\nu(m/f)^{-\frac{3}{2}}\log K \quad (m < f). \quad (3.4.24)$$

To proceed further we now replace (3.4.8) by

$$K^{\frac{1}{2}}(\log K)^2 \le G < K^{1-\delta},$$

and recall that we have $f \ll K$. Then the right sides of (3.4.20)–(3.4.21) are simplified into $O(m^{-1}\exp(-\frac{1}{5}(\log K)^2))$, and we see that the contribution of each $\mathscr{H}_j(f;h_\nu)$, $j = 2,3,5,6$, to $\mathscr{H}(f;h_\nu)$ is $O(\exp(-\frac{1}{6}\times(\log K)^2))$, which is negligible. As to $\mathscr{H}_4(f;h_\nu)$, we use (3.4.24). On the other hand, to $\mathscr{H}_7(f;h_\nu)$ we apply the classical bounds. In this way we get

$$\mathscr{H}(f;h_\nu) = \frac{2}{i\pi^3}\left\{(c_E - \log(2\pi\sqrt{f}))(\widehat{h_\nu})'(\tfrac{1}{2}) + \tfrac{1}{4}(\widehat{h_\nu})''(\tfrac{1}{2})\right\}d(f)f^{-\frac{1}{2}}$$
$$+ O\Big\{f^{\frac{3}{2}}K^3G^{-3}\log K\sum_{m<f} m^{-2}d(m)d(f-m)\Big\}$$
$$+ O\{d(f)K^3(\log K)^6\} \quad (3.4.25)$$

uniformly for $f \ll K$ and $\nu \le N_1$. The formulas (3.3.37) and (3.3.38) give

$$(\widehat{h_\nu})'(\tfrac{1}{2}) \ll K^3G(G/K)^\nu, \quad (\widehat{h_\nu})''(\tfrac{1}{2}) \ll K^3G(G/K)^\nu \log K$$

as well as

$$(\widehat{h_0})'(\tfrac{1}{2}) = 2i\pi^{\frac{3}{2}}K^3G + O(KG^3),$$
$$(\widehat{h_0})''(\tfrac{1}{2}) = 8i\pi^{\frac{3}{2}}K^3G\log K + O(KG^3\log K).$$

We insert (3.4.25) together with these into (3.4.18) while noting that $U_\nu(x) = O(1)$ uniformly for $\log x \ll \log K$ and that

$$\sum_{f\le 3K} f\sum_{m<f} m^{-2}d(m)d(f-m) \ll K^2\log K.$$

Thus we have

$$\mathscr{C}(K,G) = 4\pi^{-\frac{3}{2}}K^3G\{\mathscr{C}_1^*(K,G) + \mathscr{C}_2^*(K,G)\} + O(K^{\frac{7}{2}}(\log K)^7 + K^2G^2(\log K)^3), \qquad (3.4.26)$$

where

$$\mathscr{C}_1^*(K,G) = \sum_{f=1}^{\infty} f^{-1}d(f)(\log K + c_E - \log(2\pi\sqrt{f}))\exp(-(f/K)^\lambda)$$

and

$$\mathscr{C}_2^*(K,G) = -\sum_{f=1}^{\infty} f^{-1}d(f)(\log K + c_E - \log(2\pi\sqrt{f}))U_0(fK).$$

We have

$$\mathscr{C}_1^*(K,G) = \frac{1}{2\pi i\lambda}\int_{(1)} \{(\log K + c_E - \log(2\pi))\zeta^2(w+1) + \zeta'(w+1)\zeta(w+1)\}K^w\Gamma(w/\lambda)dw.$$

Thus, shifting the path to $\operatorname{Re} w = -1$, we get

$$\mathscr{C}_1^*(K,G) = \tfrac{1}{3}\log^3 K + O(\log^2 K).$$

Similarly we have, by (3.4.10),

$$\begin{aligned}\mathscr{C}_2^*(K,G) &= -\frac{1}{2\pi i\lambda}\int_{(-1)} \{(\log K + c_E - \log(2\pi))\zeta^2(1-w) \\ &\quad + \zeta'(1-w)\zeta(1-w)\}(4\pi^2K^{-1})^w u_0(w)\Gamma(w/\lambda)dw \\ &= \tfrac{1}{3}\log^3 K + O(\log^2 K).\end{aligned}$$

Hence we have proved that if

$$K^{\frac{1}{2}}(\log K)^5 \le G \le K^{1-\delta} \quad (0 < \delta < \tfrac{1}{2})$$

then we have, by (3.4.26),

$$\mathscr{C}(K,G) = \tfrac{8}{3}\pi^{-\frac{3}{2}}K^3G\log^3 K\big(1 + O((\log K)^{-1})\big),$$

where the implied constant may depend on δ. This gives rise to the assertion of Theorem 3.2.

3.5 Spectral large sieve

Our next problem is to estimate the spectral fourth power moment of $H_j(\frac{1}{2})$. There are at least two ways to take. One is to extend the above argument to the expression

$$\sum_{\kappa_j \le K} \alpha_j H_j(\tfrac{1}{2})^2 \Big| \sum_{n \le N} a_n t_j(n) \Big|^2.$$

The other is to appeal to a more fundamental implement due to H. Iwaniec, which we call the spectral large sieve method. We shall take the latter argument, since it provides us with a wider scope.

We begin with the classical additive large sieve inequality:

Lemma 3.10 *Let $\{x_j\}$ be a set of points such that $\min_{n\in\mathbb{Z}} |x_j - x_k - n| \ge \delta$ $(j \ne k)$, where $0 < \delta < 1$ is arbitrary. Then we have, for any complex vector $\{a_n\}$ and integers $M, N > 0$,*

$$\sum_j \Big| \sum_{N \le n < N+M} a_n e(nx_j) \Big|^2 \le (M + 1 + 2\delta^{-1}) \|a_n\|^2, \tag{3.5.1}$$

where $\|a_n\|$ is the Euclidean norm of the vector.

Proof According to a well-known duality principle in linear algebra, it is sufficient to show, for any complex vector $\{b_j\}$,

$$\sum_{N \le n < N+M} \Big| \sum_j b_j e(nx_j) \Big|^2 \le (M + 1 + 2\delta^{-1}) \|b_j\|^2. \tag{3.5.2}$$

This is equivalent to

$$S = \sum_{-N \le n \le N} \Big| \sum_j b_j e(nx_j) \Big|^2 \le (2N + 1 + 2\delta^{-1}) \|b_j\|^2.$$

To prove the latter, let $D > 0$ be an integer to be fixed later, and put

$$S^* = \sum_n w(n) \Big| \sum_j b_j e(nx_j) \Big|^2,$$

where $w(n) = 1$ if $|n| \le N$, $= (N + D - |n|)/D$ if $N < |n| \le N + D$, and $= 0$ otherwise. We have obviously

$$S \le S^*.$$

Expanding out the square we get

$$S^* = \sum_{j,k} b_j \overline{b_k} W(x_j - x_k) \le \Big(W(0) + \max_l \sum_{k \ne l} |W(x_k - x_l)| \Big) \|b_j\|^2,$$

where

$$W(x)=\sum_n w(n)e(nx).$$

We have

$$W(x)=\frac{1}{D}\Big[\Big(\frac{\sin(\pi(N+D)x)}{\sin(\pi x)}\Big)^2-\Big(\frac{\sin(\pi Nx)}{\sin(\pi x)}\Big)^2\Big].$$

Thus we have

$$W(0)=2N+D$$

and

$$|W(x)|\le (4D\min_{n\in\mathbb{Z}}|x-n|^2)^{-1}.$$

The latter estimate readily implies

$$\sum_{k\neq l}|W(x_k-x_l)|<\frac{1}{4D}\sum_{n=1}^{\infty}\frac{2}{(n\delta)^2}=\frac{\pi^2}{12D\delta^2}$$

uniformly in l. Putting $D=[\delta^{-1}]+1$ we end the proof of the lemma.

We next introduce the hybrid large sieve:

Lemma 3.11 *Let $\{x_j\}$ be as in the previous lemma. Then we have, for any $T\ge 1$,*

$$\sum_j\int_{-T}^{T}\Big|\sum_n a_n e(nx_j)n^{it}\Big|^2dt\ll\sum_n(n+T\delta^{-1})|a_n|^2, \tag{3.5.3}$$

provided the right side converges, where the implied constant is absolute.

Proof We consider the mean square of a generic trigonometric series

$$F(t)=\sum_\omega c_\omega e^{i\omega t}.$$

Let $\tau>0$ be arbitrary, and put $v(x)=\tau^{-1}$ if $|x|\le\frac{1}{2}\tau$, and $=0$ otherwise. Then, applying the ordinary Parseval formula to the Fourier transform of the convolution

$$C(x)=\sum_\omega c_\omega v(x-\omega)=\tau^{-1}\sum_{|\omega-x|\le\frac{1}{2}\tau}c_\omega,$$

we have

$$\int_{-\infty}^{\infty}|C(x)|^2dx=(2\pi)^{-1}\int_{-\infty}^{\infty}|F(t)\hat{v}(t)|^2dt.$$

provided they converge. Since

$$\hat{v}(t) = \frac{2}{\tau t}\sin(\tfrac{1}{2}\tau t),$$

which is not less than $2\pi^{-1}$ for $|\tau t| \le \pi$, we obtain

$$\int_{-\pi/\tau}^{\pi/\tau} |F(t)|^2 dt \ll \tau^{-2} \int_{-\infty}^{\infty} \Big| \sum_{|\omega - x| \le \frac{1}{2}\tau} c_\omega \Big|^2 dx.$$

We put $\tau = \pi/T$ and $c_\omega = a_n e(nx_j)$ with $\omega = \log n$, getting

$$\int_{-T}^{T} \Big| \sum_n a_n e(nx_j) n^{it} \Big|^2 dt \ll T^2 \int_0^\infty \Big| \sum_{y \le n \le y \exp(\pi/T)} a_n e(nx_j) \Big|^2 \frac{dy}{y}.$$

Combined with (3.5.1) this yields the assertion of the lemma.

Next we shall show analytical estimates which enables us to refine the argument employed in the proof of Lemma 2.4: We put

$$J(\eta; K, G) = \sin(\pi\eta) \int_{-\infty}^{\infty} \Gamma(\eta + it)\Gamma(\eta - it) \exp(-((t-K)/G)^2) dt$$

$$(\eta = \alpha + iu,\ \alpha > 0).$$

Here we assume that K is sufficiently large, and

$$C^2(\log K)^{\frac{1}{2}} \le G \le C^{-2} K (\log K)^{-\frac{1}{2}} \tag{3.5.4}$$

with a large constant $C > 0$. Then we have

$$J(\eta; K, G) \ll \begin{cases} K^{-C} & \text{if } |u| \le C^{-2} G K (\log K)^{-\frac{1}{2}}, \\ G|u|^{2\alpha - 1} & \text{if } |u| \ge C^{-2} G K (\log K)^{-\frac{1}{2}}. \end{cases} \tag{3.5.5}$$

Here the implied constants depend on α and C, but they can be taken to be absolute since in the later application α is close to $\frac{1}{2}$ and C fixed. The same remark applies to all estimates below within the present section. In (3.5.5) the first bound for $|u| \le \frac{1}{2}K$ and the second bound throughout the indicated range are trivial consequences of Stirling's formula. To deal with the remaining case we may naturally suppose that u is positive. Thus let us assume that we have

$$\tfrac{1}{2}K \le u \le C^{-2} G K (\log K)^{-\frac{1}{2}}. \tag{3.5.6}$$

Using (2.3.11), we get

$$J(\eta; K, G) = \sqrt{\pi} G \sin(\pi\eta)\Gamma(2\eta) \int_0^1 R(y) dy, \tag{3.5.7}$$

where

$$R(y) = y^{\eta+iK-1}(1-y)^{\eta-iK-1}\exp\Big(-\tfrac{1}{4}\Big(G\log\frac{y}{1-y}\Big)^2\Big).$$

It is easy to estimate the part of the integral corresponding to $|y-\frac{1}{2}|\ge r_0$ where $r_0 = CG^{-1}(\log K)^{\frac{1}{2}}$. We then complete the segment $[\frac{1}{2}-r_0, \frac{1}{2}+r_0]$ to the parallelogram contour, a side of which is $\ell_0 = [\frac{1}{2}-r_0(1-\vartheta), \frac{1}{2}+r_0(1+\vartheta)]$, where $\vartheta = \frac{1}{4}\exp(\frac{3}{4}\pi i)$. We let ℓ_- and ℓ_+ be the left and the right sides, respectively. We get

$$\int_0^1 R(y)dy = \int_{\ell_-+\ell_0+\ell_+} R(y)dy + O(K^{-C}). \tag{3.5.8}$$

To estimate the integral over ℓ_+ we put $y = \frac{1}{2}+r_0(1+\vartheta\xi)$. We have

$$\int_{\ell_-} R(y)dy \ll r_0 \int_0^1 |\exp(h_+(\xi))|d\xi$$

with the implied constant depending only on α, where

$$\begin{aligned} h_+(\xi) = {} & i(K+u)\log(1+2r_0\vartheta\xi/(1+2r_0)) \\ & - i(K-u)\log(1-2r_0\vartheta\xi/(1-2r_0)) \\ & - \tfrac{1}{4}\big\{G\log[1+4r_0(1+\vartheta\xi)/(1-2r_0(1+\vartheta\xi))]\big\}^2. \end{aligned}$$

A simple approximation argument gives

$$h_+(\xi) = 4Kr_0\xi(i\vartheta + O(ur_0K^{-1})) - 4(Gr_0(1+\vartheta\xi))^2(1+O(r_0)).$$

Thus we have, by (3.5.6),

$$\operatorname{Re} h_+(\xi) \le -\tfrac{1}{2}(Gr_0)^2,$$

which obviously ends the treatment of the segment ℓ_+. The same argument applies to ℓ_-. As to ℓ_0, we put $y = \frac{1}{2}+r_0\vartheta + r$, and get

$$\int_{\ell_0} R(y)dy \ll \int_{-r_0}^{r_0} |\exp(h(r,u))|dr$$

with the implied constant depending only on α, where

$$\begin{aligned} h(r,u) = {} & i(K+u)\log(1+2r_0\vartheta/(1+2r)) \\ & - i(K-u)\log(1-2r_0\vartheta/(1-2r)) \\ & - \tfrac{1}{4}\big\{G\log[1+4(r+r_0\vartheta)/(1-2(r+r_0\vartheta))]\big\}^2. \end{aligned}$$

We have

$$\begin{aligned} h(r,u) = {} & 4ir_0\vartheta K(1-2ruK^{-1}) \\ & - 4ur_0^2(1+O(r_0)) - 4(G(r+r_0\vartheta))^2(1+O(r_0)); \end{aligned}$$

and thus, by (3.5.4) and (3.5.6),

$$\operatorname{Re} h(r,u) < -\tfrac{1}{2}K r_0. \tag{3.5.9}$$

This ends the proof of (3.5.5).

We shall also need the following estimate: Let us put

$$P(x;K,G) = (2\pi^2 i)^{-1} \int_{(\alpha)} \eta^{-1} J(\eta;K,G)\Big(\frac{x}{2}\Big)^{1-2\eta} d\eta \quad (0<\alpha<\tfrac{1}{2}). \tag{3.5.10}$$

Then we have

$$P(x;K,G) \ll x K^{-C}, \tag{3.5.11}$$

provided (3.5.4) holds and

$$0 < x \le C^{-2} K G (\log K)^{-\frac{1}{2}}. \tag{3.5.12}$$

This is to be compared with (2.3.13). The proof starts with the representation

$$P(x;K,G) = -\frac{Gx}{2\sqrt{\pi}} \int_0^1 y^{iK-1}(1-y)^{-iK-1} \operatorname{si}\Big(\frac{x}{2\sqrt{y(1-y)}}\Big) \times \exp\Big(-\tfrac{1}{4}\Big(G\log\frac{y}{1-y}\Big)^2\Big) dy,$$

which corresponds to (2.3.12) and can be shown in just the same way. We have

$$P(x;K,G) = -\frac{Gx}{2\sqrt{\pi}} \int_{\ell_- + \ell_0 + \ell_+} \cdots dy + O(xK^{-C}),$$

where the path is as in (3.5.8), and (3.5.12) is supposed; note that the si-factor is regular in the vicinity of $y=\frac{1}{2}$. To estimate the right side we shall deal with only the segment ℓ_0, since the others are as before. We have

$$\int_{\ell_0} \cdots dy \ll \int_{-r_0}^{r_0} |\exp(h(r,0))||\operatorname{si}(xb(r))| dr,$$

where $b(r) = (1-4(r+r_0\vartheta)^2)^{-\frac{1}{2}}$. The estimate (3.5.9) holds for $h(r,0)$, too. On the other hand we have

$$\operatorname{si}(xb(r)) = \operatorname{si}(x) + \int_x^{xb(r)} \frac{\sin\xi}{\xi} d\xi \ll 1 + x^{-1}\int_x^{xb(r)} \exp(|\operatorname{Im}\xi|)|d\xi| \ll \exp(8r_0^2 x).$$

Hence the contribution of the segment ℓ_0 to $P(x;K,G)$ is negligible, provided (3.5.12) holds. This ends the proof of (3.5.11).

Now we shall prove the following version of the spectral large sieve inequality:

Theorem 3.3 *Let $t_j(n)$ be the Hecke eigenvalue defined by* (3.1.13), *and let α_j be as in* (3.3.1). *Further let us suppose* (3.5.4). *Then the inequality*

$$\sum_{K \le \kappa_j \le K+G} \alpha_j \Big| \sum_{N \le n \le 2N} a_n t_j(n) \Big|^2 \ll (KG + N(\log K)^{\frac{3}{2}}) \|a_n\|^2 \tag{3.5.13}$$

holds for any complex vector $\{a_n\}$, where $\|a_n\|$ is the Euclidean norm of the vector. Here the implied constant depends only on C involved in (3.5.4).

Proof We shall first deal with the case where $N \le K^B$ with an arbitrary fixed $B > 0$: We use (2.3.4) with the present convention that $\rho_j(n) = \rho_j(1)t_j(n)$. We multiply both sides of (2.3.4) by the factor

$$K\overline{a_m}a_n t^{-1}\sinh(\pi t)\exp(-((t-K)/G)^2),$$

integrate over the real axis with respect to t, and sum over $N \le m,\, n \le 2N$. We may discard the contribution of the continuous spectrum because it is non-negative. Then, corresponding to (2.3.10), we get

$$\sum_{K \le \kappa_j \le K+G} \alpha_j \Big| \sum_{N \le n \le 2N} a_n t_j(n) \Big|^2 \ll KG\|a_n\|^2$$
$$+K\Big| \sum_{N \le m,n \le 2N} \overline{a_m}a_n \sum_{l=1}^{\infty} l^{-1}S(m,n;l)P\left(\frac{4\pi}{l}\sqrt{mn};K,G\right)\Big|,$$

where the function P is defined by (3.5.10). Thus we see, by (3.5.11), that the terms with $l \ge 4\pi C^2 N(KG)^{-1}(\log K)^{\frac{1}{2}}$ contribute

$$\ll NK^{1-C} \sum_{N \le n \le 2N} |a_n|^2 \sum_{l=1}^{\infty} l^{-2} \sum_{N \le m \le 2N} |S(m,n;l)|$$
$$\ll K^{3B-C}\|a_n\|^2$$

because of (2.1.4). If $N \le (4\pi C^2)^{-1}KG(\log K)^{-\frac{1}{2}}$ then this ends the proof of (3.5.13). Thus let us assume that $N > (4\pi C^2)^{-1}KG(\log K)^{-\frac{1}{2}}$. We have

$$\sum_{K \le \kappa_j \le K+G} \alpha_j \Big| \sum_{N \le n \le 2N} a_n t_j(n) \Big|^2 \ll KG\|a_n\|^2 + K\sum_{l \le l_0} l^{-1}|Y(l)|, \tag{3.5.14}$$

where $l_0 = 4\pi C^2 N(KG)^{-1}(\log K)^{\frac{1}{2}}$, and

$$Y(l) = \sum_{N \le m,n \le 2N} \overline{a_m} a_n S(m,n;l) P\left(\frac{4\pi}{l}\sqrt{mn}; K, G\right).$$

Replacing the Kloosterman sum and the function P by their defining expressions, we get

$$Y(l) \ll l^{2\alpha-1} \sum_{\substack{h=1 \\ (h,l)=1}}^{l} \int_{(\alpha)} |\eta^{-1} J(\eta; K, G)| \Big| \sum_{N \le n \le 2N} a_n e(hn/l) n^{\frac{1}{2}-\eta} \Big|^2 |d\eta|.$$

We note that this integral converges if $\alpha < \frac{1}{2}$. We keep α close to $\frac{1}{2}$. Using (3.5.5) we have

$$\begin{aligned} Y(l) &\ll K^{2B-C} \|a_n\|^2 \\ &+ G l^{2\alpha-1} \sum_{\substack{h=1 \\ (h,l)=1}}^{l} \int_{u_0}^{\infty} u^{2(\alpha-1)} \Big| \sum_{N \le n \le 2N} a_n e(hn/l) n^{\frac{1}{2}-\alpha+iu} \Big|^2 du, \end{aligned}$$

where $u_0 = C^{-2} KG(\log K)^{-\frac{1}{2}}$. Thus we have, by (3.5.3),

$$\begin{aligned} \sum_{l \le l_0} l^{-1} |Y(l)| &\ll K^{2B-C+1} \|a_n\|^2 \\ &+ G N^{1-2\alpha} \sum_{L,U} (LU)^{2(\alpha-1)} (N + UL^2) \|a_n\|^2, \end{aligned}$$

where L and U run over the sequence of integral powers of 2 less than l_0 and larger than u_0, respectively. Hence we get

$$\begin{aligned} &\sum_{l \le l_0} l^{-1} |Y(l)| \ll K^{2B-C+1} \|a_n\|^2 \\ &+ \left[(\tfrac{1}{2} - \alpha)^{-1} N K^{-1} (\log K)^{\frac{1}{2}} + G(NK^{-1}G^{-1}(\log K)^{\frac{1}{2}})^{2(1-\alpha)} \right] \|a_n\|^2. \end{aligned}$$

In this we put $\alpha = \frac{1}{2} - (\log K)^{-1}$, and insert the result into (3.5.14), finishing the proof of (3.5.13), provided N is less than a fixed power of K.

We then consider the case where N is greater than a fixed power of K. This time we use the duality principle (cf. (3.5.2)). Thus we consider, instead, the estimation of the expression

$$\sum_{N \le n \le 2N} \Big| \sum_{K \le \kappa_j \le K+G} (\cosh \pi \kappa_j)^{-\frac{1}{2}} \rho_j(n) b_j \Big|^2. \tag{3.5.15}$$

Here $\{b_j\}$ is an arbitrary complex vector; and we have used (3.1.16) and

(3.3.1). Let $\omega(x)$ be a C^∞-function which is equal to 1 for $1 \le x \le 2$, and equal to 0 for $x \le \frac{1}{2}$ or $x \ge \frac{5}{2}$. Then it is sufficient to bound

$$\sum_n \omega(n/N) \Big| \sum_j (\cosh \pi\kappa_j)^{-\frac{1}{2}} \rho_j(n) b_j \Big|^2,$$

where $b_j = 0$ if $\kappa_j > 2K$. Applying the Mellin inversion formula to ω and expanding out the square, we see that the above is equal to

$$\frac{1}{2\pi i} \sum_{j,j'} \frac{b_j \overline{b_{j'}}}{(\cosh \pi\kappa_j \cosh \pi\kappa_{j'})^{\frac{1}{2}}} \int_{(\alpha)} R_{j,j'}(s)\hat{\omega}(s) N^s ds \quad (\alpha > 1), \tag{3.5.16}$$

where $R_{j,j'}$ is defined by (3.2.3), and $\hat{\omega}$ is the Mellin transform of ω, which decays faster than any fixed negative power of $|s|$ as $|\mathrm{Im}\, s|$ increases. Then, by virtue of Lemma 3.5, especially by the estimate (3.2.13), we have

$$\int_{(\alpha)} R_{j,j'}(s)\hat{\omega}(s) N^s ds = 24\pi^{-1} i \delta_{j,j'} \hat{\omega}(1)(\cosh \pi\kappa_j) N + O(|\rho_j(1)\rho_{j'}(1)| K^c N^{\frac{3}{4}}).$$

Inserting this into (3.5.16) and invoking (3.3.2), we find that if $N \ge K^B$ then (3.5.15) is

$$\ll N \|b_j\|^2 \tag{3.5.17}$$

with the implied constant depending only on B. This ends the proof of Theorem 3.3.

We are now ready to prove the crucial bound:

Theorem 3.4 *We have*

$$\sum_{\kappa_j \le K} \alpha_j H_j(\tfrac{1}{2})^4 \ll K^2 (\log K)^{15}. \tag{3.5.18}$$

Proof We consider first the part of the sum where $\frac{1}{2}K \le \kappa_j \le K$. We note that (3.4.9) gives, uniformly for such κ_j's,

$$H_j(\tfrac{1}{2}) \ll 1 + \log K \sum_{\pm} \int_{(\pm\lambda^{-1})} \Big| \sum_{n \le 3K} t_j(n) n^{-\frac{1}{2} \mp w} \Big| \exp(-|w|/\lambda) |dw|,$$

where $\lambda = c_0 \log K$ with an appropriately chosen $c_0 > 0$. Thus we have, by Hölder's inequality,

$$H_j(\tfrac{1}{2})^4 \ll 1 + (\log K)^7 \sum_{\pm} \int_{(\pm\lambda^{-1})} |Q_j(\mp w)|^2 \exp(-|w|/\lambda) |dw|, \tag{3.5.19}$$

where

$$Q_j(w) = \{ \sum_{n \le 3K} t_j(n) n^{-\frac{1}{2}+w} \}^2.$$

The relation (3.1.14) gives

$$Q_j(w) \ll \sum_{l \le 3K} l^{-1} \Big| \sum_{n \le 9(K/l)^2} d^*(n) t_j(n) n^{-\frac{1}{2}+w} \Big|;$$

and thus

$$Q_j(w)^2 \ll (\log K)^2 \sum_{l \le 3K} l^{-1} \sum_N \Big| \sum_{N \le n \le 2N} d^*(n) t_j(n) n^{-\frac{1}{2}+w} \Big|^2.$$

Here $|\mathrm{Re}\, w| \ll (\log K)^{-1}$, and $N \le 9(K/l)^2$ runs over the sequence of integral powers of 2. Further, $d^*(n)$ is the number of ways of representing n as a product of two integers $\le 3K/l$. Then, dividing the range $[\frac{1}{2}K, K]$ into intervals of length $\approx K(\log K)^{-\frac{1}{2}}$ we get, by (3.5.13),

$$\sum_{\frac{1}{2}K \le \kappa_j \le K} \alpha_j |Q_j(w)|^2 \ll (\log K)^2 \sum_{l \le 3K} l^{-1} \sum_N (K^2 + N(\log K)^2)(\log N)^3$$
$$\ll K^2 (\log K)^7$$

uniformly for w such that $|\mathrm{Re}\, w| \ll (\log K)^{-1}$. From this and (3.3.2), (3.5.19), we obtain (3.5.18).

3.6 Notes for Chapter 3

Prior to Hecke's far-reaching investigations [17, p. 577 to the end] Mordell had used the operators $T_6(p)$ for primes p in his proof of Ramanujan's conjecture on the multiplicativity of the Fourier coefficients of the modular discriminant,

$$\eta(z) = e^{2\pi i z} \prod_{n=1}^{\infty} (1 - e^{2\pi i n z})^{24},$$

which is a cusp-form of weight 12 and an eigenfunction of all Hecke operators $T_6(n)$ (see [16, Chapter X], [17, p. 670]). We note also that the operators $T_6(p)$ occurred first in a work of Hurwitz. Nevertheless, it is appropriate to attribute these operators to Hecke. For it was Hecke who grasped the concept as a structural element in the entire theory of automorphic functions. Naturally the theory takes the simplest form in the case of the full modular group. Already Hecke [17, p. 672] considered the case of congruence subgroups, where he encountered difficulties in fully extending his findings about the multiplicativity of Fourier coefficients of modular forms. This was later resolved by Atkin and Lehner [1]. For the general theory of Hecke operators see, e.g., Rankin [61, Chapter 9].

Hecke's theory was greatly enhanced by Petersson's metric theory [59]. Lemmas 3.2 and 3.4 are due to Maass [41], and are extensions to real-analytic cusp-forms of the Hecke–Petersson theory given briefly in Lemma 3.6. As far as the action of Hecke operators is concerned, there are no specific differences between the holomorphic and real-analytic forms. Thus the structural theory of Hecke operators can be applied to both types of automorphic forms. There exists, however, a notable difference, too. That is about the size of Hecke eigenvalues: Ramanujan conjectured empirically that in the case of the function $\eta(z)$ the exponent $\frac{1}{4}$ in (3.1.22) could be replaced by any positive constant; and this was extended to a general situation by Petersson. Later Deligne proved their conjecture in the stronger form

$$|t_{j,k}(n)| \le d(n) \quad (j, k \ge 1),$$

and in fact more. The analogue of this for real-analytic cusp-forms has not been established yet, though it is considered highly plausible. The hitherto best result on this subject is

$$|t_j(n)| \le d(n) n^{\frac{5}{28}} \quad (j \ge 1)$$

due to Bump, Duke, Gintzberg, Hoffstein and Iwaniec [6]. So far no evidences have been obtained which suggest any relation between the zeta-function and the Ramanujan–Petersson conjecture. The bounds (3.1.18) and (3.1.22), which depend on (2.1.4), are more than enough for our later purpose. In fact, any bound suffices if it can yield (3.2.5), (3.2.13), and (3.2.22). The size of Hecke L-functions corresponding to Maass forms, especially on the critical line $\mathrm{Re}\, s = \frac{1}{2}$, is an important issue. For this see Iwaniec [26], Meurman [42], and Sarnak [64].

Rankin (see [61, p. 174]) introduced his L-function to study the above conjecture of Ramanujan. It should be noted that as a matter of fact we need only the case $\varepsilon_j \varepsilon_{j'} = +1$ in our application of Lemma 3.5, i.e., in the proof of Theorem 3.3. The Rankin L-functions for holomorphic cusp-forms as well as for convolutions of holomorphic and real-analytic cusp-forms can be treated similarly. It is then easy to show that

$$\sum_{n \le N} |a(n)|^2 \ll N,$$

where $a(n)$ stands for the generic Fourier coefficient of those cusp forms (cf. (3.5.17)). Thus, in particular, the defining series for $H_j(s)$, $H_{j,k}(s)$ are absolutely convergent for $\mathrm{Re}\, s > 1$.

The Hecke correspondence between automorphic forms and Dirichlet series, especially those with Euler products, has been a main theme in the theory of automorphic functions. We are not in a position to dwell on this fascinating subject. But we should point out that our main result (Theorem 4.2) in the next chapter can be viewed in this perspective, too. It relates the zeta-function with cusp-forms over the full modular group, though this time the correspondence is between the single zeta-function and infinitely many forms. Thus it can be said in other words that spectro-statistical estimates of Hecke L-functions would probably be more important than estimates of individual values of each L-function. This is the reason that we developed such a theory in the above. The results in Sections 3.3–3.4 are due to Motohashi [46] (cf. Kuznetsov [37]). We note that the inequality (3.4.3) is best possible, given the relevant conditions.

The identity (3.3.33), which was claimed by Kuznetsov [37] without rigorous proof, illustrates the importance of the binary additive divisor problems

$$\sum_{n\le N} d(n)d(n+f), \quad \sum_{n<N} d(n)d(N-n),$$

where $f \ge 1$. This subject has a rich history. For a recent account see Motohashi [48]. Because of the expansion (1.1.9), the above sums are related with the convolution of the Eisenstein series. From such a viewpoint Vinogradov and Takhtadjan [73] developed an interesting argument. Theirs does not depend on the specific structure of $d(n)$ that is essential in Kuznetsov and Motohashi's method. Thus their argument is capable of extension to convolution of cusp-forms. In the case of holomorphic cusp-forms this approach had been investigated and applied to the mean-value problem of Hecke L-functions by Good ([13], [14]) (see also Motohashi [50]). On the other hand the case of real-analytic cusp-forms is the subject of Jutila's investigation [28].

We could make the argument of the proof of Theorem 3.1 more sophisticated so that the bounds are replaced by asymptotic expressions.

For example we can show that

$$\sum_{\kappa_j \le K} \alpha_j H(\tfrac{1}{2})^2 = 2\pi^{-2}K^2(\log K + c_E - \tfrac{1}{2} - \log(2\pi)) + O(K(\log K)^6)$$

(see Motohashi [46]; Kuznetsov's relevant claims in [37] are incorrect). This implies of course the infinitude of the cardinality of the set $\{\kappa_j\}$, which was already proved at (2.3.14). More interesting is that this asymptotic formula implies (3.4.17), too. The deduction of this fact is, however, dependent on a far deeper result due to Katok and Sarnak [34]. Using an extension of Shimura's theory of holomorphic cusp-forms of half-integral weights, they could show that each $H_j(\frac{1}{2})$ is non-negative, which immediately yields the assertion (3.4.17). Needless to say, our proof of (3.4.17) is definitely simpler. But it is worth stressing that such an arithmetic fact as $H_j(\frac{1}{2}) \ge 0$ has an important implication in the theory of the zeta-function (see Theorem 5.5 below).

Iwaniec [22] observed, on the basis of Kuznetsov's treatment [36, Section 5] of the spectral mean square of Maass–Fourier coefficients, that these coefficients behave like ordinary characters on average. He then promptly created the spectral large sieve method. The wealth of fascinating applications of this versatile method can be found in his surveys [24], [25] (modulo a premature claim on the additive divisor problem). A bound slightly weaker than (3.5.18) was proved by Motohashi [46] with an extension of the argument of Section 3.4; thus we could dispense with the spectral large sieve method. We think, however, that the proof of (3.5.18) as it is developed there would be an especially suitable opportunity to indicate the essentials of Iwaniec's method. Theorem 3.3 is an improved version of an inequality of Iwaniec [22, Theorem 1]. It should be observed that (3.5.13) rests on the three fundamental methods in modern number theory: Bombieri's large sieve (Lemma 3.10), Gallagher's hybrid large sieve (Lemma 3.11), and the Selberg–Kuznetsov theory of Kloosterman sums. Our proof of (3.5.13) is modeled on Jutila [31], and should be compared with Iwaniec [22]. We have avoided involved formulas related to Bessel functions, which were used by Kuznetsov and Iwaniec. This simplification was made possible by the separation of variables achieved in (2.1.2). From our argument it can be seen that the spectral large sieve is essentially a consequence of the Linnik–Bombieri large sieve. The way connecting these sieves is the trace formula (2.3.16) of Kuznetsov.

Iwaniec considered also the hybrid spectral mean ([22, Theorem 2])

$$\sum_{\kappa_j \le K} \alpha_j \Big| \sum_{n \le N} a_n t_j(n) \Big|^2 \Big| \sum_{r \le R} b_r r^{i\kappa_j} \Big|^2 .$$

The argument of the last section can be extended so as to include this expression, too. The result (3.4.3) is obviously a hybrid type, but is more suitable to our later application (see Theorem 5.3 below). See also Jutila [31] and the references given there.

Here are some additional remarks: One may wonder about the size of $\rho_j(1)$. For this see Iwaniec [27, pp. 130–132]. As to the classical large sieve method see Linnik [39] and Bombieri [3]; and for sieve methods in general see, e.g., Motohashi [44].

4
An explicit formula

We are now ready to begin our investigation of the Riemann zeta-function. The functional equation (1.1.13) implies that most secrets of $\zeta(s)$, $s = \sigma + it$, are buried in the critical strip $0 \le \sigma \le 1$. The history of the study of the zeta-function can be summarized as the series of impressive efforts to dig out those secrets and feed them back to various classical problems in number theory. Especially the issue of the order of magnitude of $\zeta(s)$ in the critical strip or rather on the critical line $\sigma = \frac{1}{2}$ is of profound importance in applications; but it is also a problem of extraordinary difficulty. We are, however, blessed with the fact that in many important problems, including those on the distribution of prime numbers, it is not imperative to have good bounds for individual values of $\zeta(s)$ but it is enough to have corresponding statistical estimates, i.e., results on various mean values of $\zeta(s)$. Hereby emerges the theory of the power moments

$$\mathscr{Z}_k(g) = \int_{-\infty}^{\infty} |\zeta(\tfrac{1}{2} + it)|^{2k} g(t)dt,$$

where k is a positive integer and the function g is of rapid decay. Even though only the initial cases $k = 1, 2$ have been successfully investigated and the cases $k \ge 3$ have remained virtually untouched so far, the theory of the power moments has always been at the core of analytic number theory as an indispensable tool. Moreover, the significance of the theory of $\mathscr{Z}_k(g)$ is not limited to its sheer applications to classical problems. In fact, already in the case $k = 2$, the fourth power moment, can be regarded as a means, essentially unique at present, which connects the theory of $\zeta(s)$ with the rich theory of real-analytic automorphic forms and thus allows us to gain a greater perspective on $\zeta(s)$.

It is possible to develop a theory that allows us to treat the zeta- and allied functions simultaneously, accomplishing a considerable generality. We think, however, that such a general argument would eventually fail to touch the exquisite characteristics which are very special to $\zeta(s)$. Hence we shall positively exploit the advantage offered by the peculiarities of $\zeta(s)$ and leave any possible generalizations for readers. Nevertheless there is a technical point where we have to try to be general in view of future applications, which is the choice of the weight function g. For this purpose we introduce the

Basic assumption: *The function $g(r)$ takes real values on the real axis; and there exists a large positive constant A such that $g(r)$ is regular and $O((|r|+1)^{-A})$ in the horizontal strip $|\mathrm{Im}\, r| \le A$.*

This is to avoid unnecessary complexities while attaining a modest generality. The symbol A is to be used in this context throughout the sequel. The implied constants are possibly dependent on A.

4.1 A prototype

In this section we shall study the case $k = 1$ as the initial step. We shall prove an explicit formula for $\mathscr{Z}_1(g)$ which is to be generalized to the fourth power moment situation in the subsequent sections.

To this end we consider

$$\mathscr{I}(u,v;g) = \int_{-\infty}^{\infty} \zeta(u+it)\zeta(v-it)g(t)dt, \tag{4.1.1}$$

where $\mathrm{Re}\, u, \mathrm{Re}\, v > 1$. Moving the line of integration appropriately we see that $\mathscr{I}(u,v;g)$ exists as a meromorphic function in the domain

$$|u|,\ |v| < B \quad (B = cA), \tag{4.1.2}$$

where c is a small positive constant but B is supposed to be still sufficiently large. We shall assume (4.1.2) throughout the present section. Thus various conditions on u, v below are to be understood as imposed along with (4.1.2).

We have, for $\mathrm{Re}\, u, \mathrm{Re}\, v < 1$,

$$\mathscr{I}(u,v;g) = \int_{-\infty}^{\infty} \zeta(u+it)\zeta(v-it)g(t)dt \\ + 2\pi\zeta(u+v-1)\{g((u-1)i) + g((1-v)i)\}. \tag{4.1.3}$$

To obtain this we shift the line of integration in (4.1.1) to $\mathrm{Im}\, t = 2B$, passing over the pole at $t = (u-1)i$. The resulting integral is absolutely convergent and regular in (4.1.2), which is a simple consequence of our basic assumption on g. Then we impose $\mathrm{Re}\,u$, $\mathrm{Re}\,v < 1$ and shift the path back to the original, encountering the pole at $t = (1-v)i$. Taking into account the relevant residues we get (4.1.3).

In (4.1.3) we put $u = v = \alpha$ with arbitrary $0 < \alpha < 1$; and get

$$\int_{-\infty}^{\infty} |\zeta(\alpha+it)|^2 g(t)dt = \mathscr{I}(\alpha,\alpha;g) - 4\pi\zeta(2\alpha-1)\mathrm{Re}\,\{g((\alpha-1)i)\}. \quad (4.1.4)$$

Thus

$$\mathscr{Z}_1(g) = \mathscr{I}(\tfrac{1}{2},\tfrac{1}{2};g) + 2\pi\mathrm{Re}\,\{g(\tfrac{1}{2}i)\}. \quad (4.1.5)$$

This indicates that we should look for a *different* manner of analytic continuation of $\mathscr{I}(u,v;g)$ to the region containing the point $(\frac{1}{2},\frac{1}{2})$ so that we may get an expression that reveals finer structures of $\mathscr{Z}_1(g)$. For this purpose we appeal to a dissection argument: We divide $\mathscr{I}(u,v;g)$, in the region of absolute convergence, into three parts,

$$\begin{aligned}\mathscr{I}(u,v;g) &= \Big\{\sum_{m=n} + \sum_{m<n} + \sum_{m>n}\Big\} m^{-u} n^{-v} g^*(\log(n/m)) \\ &= \zeta(u+v)g^*(0) + \mathscr{I}_1(u,v;g) + \overline{\mathscr{I}_1(\overline{v},\overline{u};g)}, \end{aligned} \quad (4.1.6)$$

say, where $\mathscr{I}_1$ corresponds to the second part, and g^* is the Fourier transform

$$g^*(\xi) = \int_{-\infty}^{\infty} g(t)e^{i\xi t}dt.$$

We shall show that $\mathscr{I}_1$ can be continued meromorphically to the domain (4.1.2).

To this end we introduce the Mellin transform

$$\tilde{g}(s,\lambda) = \int_0^{\infty} y^{s-1}(1+y)^{-\lambda} g^*(\log(1+y))dy. \quad (4.1.7)$$

As a function of two complex variables $\tilde{g}(s,\lambda)$ is regular in the domain $\{0 < \mathrm{Re}\,s < \mathrm{Re}\,\lambda + A\}$, for we have

$$\tilde{g}(s,\lambda) = \int_0^{\infty} y^{s-1}(1+y)^{-\lambda} \int_{-\infty+Ai}^{\infty+Ai} g(t)(1+y)^{it}dtdy, \quad (4.1.8)$$

which is absolutely and uniformly convergent there.

Lemma 4.1 *The function $\tilde{g}(s,\lambda)/\Gamma(s)$ continues holomorphically to the domain*

$$|\mathrm{Re}\,s| \le \tfrac{1}{3}A, \quad |\mathrm{Re}\,\lambda| \le \tfrac{1}{3}A; \quad (4.1.9)$$

and there we have

$$\tilde{g}(s,\lambda) \ll |s|^{-\frac{1}{2}A}, \tag{4.1.10}$$

when s tends to infinity while λ remains bounded.

Proof Exchanging the order of integration in (4.1.8) we get

$$\tilde{g}(s,\lambda) = \Gamma(s)\int_{-\infty+Ai}^{\infty+Ai} \frac{\Gamma(\lambda - it - s)}{\Gamma(\lambda - it)} g(t)dt. \tag{4.1.11}$$

This integral is regular for $1 - A < \mathrm{Re}\, s < \mathrm{Re}\,\lambda + A$; to see this we need only to apply Stirling's formula to the integrand and invoke the decay property of g. Thus we obtain the first assertion of the lemma. As to the bound (4.1.10) we note that (4.1.8) gives, for any integer v such that $0 \le v < A - 1$,

$$\tilde{g}(s,\lambda) = \frac{1}{s(s+1)\cdots(s+v-1)}\int_0^\infty \frac{y^{s+v-1}}{(1+y)^{\lambda+v}}$$
$$\times \int_{-\infty+Ai}^{\infty+Ai} (\lambda - it)(\lambda - it + 1)\cdots(\lambda - it + v - 1)g(t)(1+y)^{it}dtdy.$$

Then the assertion follows if we put $v = [\frac{1}{2}A] + 1$, and we end the proof of the lemma.

Now, by Mellin's inversion formula we have, for any $y > 0$,

$$(1+y)^{-\lambda}g^*(\log(1+y)) = \frac{1}{2\pi i}\int_{(2)} \tilde{g}(s,\lambda)y^{-s}ds. \tag{4.1.12}$$

Thus we have, for $\mathrm{Re}\,(u+v) > 3$,

$$\begin{aligned}\mathscr{I}_1(u,v;g) &= \sum_{m,n>0} m^{-u-v}(1+n/m)^{-v}g^*(\log(1+n/m))\\ &= \frac{1}{2\pi i}\int_{(2)} \zeta(s)\zeta(u+v-s)\tilde{g}(s,v)ds;\end{aligned}$$

and we move the path to $\mathrm{Re}\, s = 3B$, getting

$$\mathscr{I}_1(u,v;g) = \zeta(u+v-1)\tilde{g}(u+v-1,v) + \frac{1}{2\pi i}\int_{(3B)} \zeta(s)\zeta(u+v-s)\tilde{g}(s,v)ds.$$

Obviously this yields a meromorphic continuation of $\mathscr{I}_1$ to the domain (4.1.2).

Invoking the functional equation (1.1.13) we have

$$\begin{aligned}\mathscr{I}_1(u,v;g) &= \zeta(u+v-1)\tilde{g}(u+v-1,v)\\ &\quad+\frac{1}{\pi i}\int_{(3B)}(2\pi)^{u+v-1-s}\sin(\tfrac{1}{2}(u+v-s)\pi)\\ &\quad\times\Gamma(s+1-u-v)\zeta(s)\zeta(s+1-u-v)\tilde{g}(s,v)ds\,;\end{aligned}$$

and thus

$$\begin{aligned}\mathscr{I}_1(u,v;g) &= \zeta(u+v-1)\tilde{g}(u+v-1,v)-2i(2\pi)^{u+v-2}\\ &\quad\times\sum_{n=1}^{\infty}\sigma_{u+v-1}(n)\int_{(3B)}(2\pi n)^{-s}\sin(\tfrac{1}{2}(u+v-s)\pi)\\ &\quad\times\Gamma(s+1-u-v)\tilde{g}(s,v)ds, \qquad (4.1.13)\end{aligned}$$

where in order to check the absolute convergence we have used Stirling's formula for the Γ-factor and Lemma 4.1 for the $\tilde{g}$-factor. Since the region (4.1.2) is symmetric, we may interchange the rôles of u and v in the last identity. Hence, from (4.1.4), (4.1.6), and (4.1.13), we obtain, for any $0<\alpha<1$,

$$\begin{aligned}&\int_{-\infty}^{\infty}|\zeta(\alpha+it)|^2 g(t)dt = \zeta(2\alpha)g^*(0)+2\zeta(2\alpha-1)\mathrm{Re}\,\{\tilde{g}(2\alpha-1,\alpha)\}\\ &-4\pi\zeta(2\alpha-1)\mathrm{Re}\,\{g((\alpha-1)i)\}+4(2\pi)^{2\alpha-2}\\ &\times\mathrm{Im}\left[\sum_{n=1}^{\infty}\sigma_{2\alpha-1}(n)\int_{(2)}(2\pi n)^{-s}\sin(\tfrac{1}{2}(2\alpha-s)\pi)\Gamma(s+1-2\alpha)\tilde{g}(s,\alpha)ds\right],\end{aligned} \tag{4.1.14}$$

where we have moved the line $\mathrm{Re}\,s=3B$ to $\mathrm{Re}\,s=2$, which is legitimate for an obvious reason.

The singularity at $\alpha=\frac{1}{2}$ of the first term on the right side of (4.1.14) is canceled out by that of the second term. To see this we note, for α close to $\frac{1}{2}$, that

$$\zeta(2\alpha-1)=-\tfrac{1}{2}-\log(2\pi)(\alpha-\tfrac{1}{2})+O(|\alpha-\tfrac{1}{2}|^2),$$

and that (4.1.11) gives

$$\begin{aligned}\tilde{g}(2\alpha-1,\alpha) &= (2\alpha-1)^{-1}\{1+\Gamma'(1)(2\alpha-1)+O(|\alpha-\tfrac{1}{2}|^2)\}\\ &\quad\times\int_{-\infty}^{\infty}\left\{1-2\frac{\Gamma'}{\Gamma}(\tfrac{1}{2}-it)(\alpha-\tfrac{1}{2})+O(|\alpha-\tfrac{1}{2}|^2)\right\}g(t)dt.\end{aligned}$$

Hence

$$\lim_{\alpha\to\frac{1}{2}}\left[\zeta(2\alpha)g^*(0)+2\zeta(2\alpha-1)\operatorname{Re}\{\tilde{g}(2\alpha-1,\alpha)\}\right]$$
$$=\int_{-\infty}^{\infty}\left[\operatorname{Re}\left\{\frac{\Gamma'}{\Gamma}(\tfrac{1}{2}+it)\right\}+2c_E-\log(2\pi)\right]g(t)dt, \qquad (4.1.15)$$

where c_E is the Euler constant.

Next, we transform

$$\int_{(2)}(2\pi n)^{-s}\sin(\tfrac{1}{2}\pi(2\alpha-s))\Gamma(s+1-2\alpha)\tilde{g}(s,\alpha)ds = P_+ - P_-,$$

where

$$P_\pm = \frac{1}{2i}\int_{(2)}(2\pi n)^{-s}e^{\pm\frac{1}{2}\pi i(2\alpha-s)}\Gamma(s+1-2\alpha)\tilde{g}(s,\alpha)ds.$$

To eliminate a difficulty about convergence we introduce

$$g_\delta(t) = e^{-\delta t^2}g(t) \quad (\delta\ge 0),$$

and denote by $P_\pm(\delta)$ the result of replacing g in $P_\pm$ by g_δ. Obviously the assertion of Lemma 4.1 holds for $\tilde{g_\delta}(s,\lambda)$ uniformly for $\delta\ge 0$. That is, we have

$$\lim_{\delta\to 0+}P_\pm(\delta) = P_\pm.$$

But we have, for $\delta > 0$,

$$P_+(\delta) = \frac{e^{\pi i\alpha}}{2i}\int_{(2)}(2\pi n)^{-s}e^{-\frac{1}{2}\pi is}\Gamma(s+1-2\alpha)$$
$$\times\int_0^{e^{i\theta}\infty}y^{s-1}(1+y)^{-\alpha}g_\delta^*(\log(1+y))dyds,$$

where $0<\theta<\frac{1}{2}\pi$. In fact the function $g_\delta^*(\log(1+y))$ is regular and $O(|y|^{-A})$ uniformly for $\operatorname{Re} y\ge 0$ as can be seen from its definition; and the double integral converges absolutely. Changing the order of integration, computing the resulting inner integral, and restoring the contour of the y-integral to the original, we get

$$P_+(\delta) = i\pi(2\pi n)^{1-2\alpha}\int_0^\infty y^{2(\alpha-1)}(1+y)^{-\alpha}e^{-2\pi in/y}g_\delta^*(\log(1+y))dy$$
$$= i\pi(2\pi n)^{1-2\alpha}\int_0^\infty (y(1+y))^{-\alpha}e^{-2\pi iny}g_\delta^*(\log(1+1/y))dy.$$

By virtue of the presence of the factor $e^{-2\pi iny}$ the last integral converges uniformly for $\delta \geq 0$, as integration by parts yields. Hence we find that

$$P_+ = i\pi(2\pi n)^{1-2\alpha} \int_0^\infty (y(1+y))^{-\alpha} e^{-2\pi i n y} g^*(\log(1+1/y))dy.$$

In just the same way we have also

$$P_- = -i\pi(2\pi n)^{1-2\alpha} \int_0^\infty (y(1+y))^{-\alpha} e^{2\pi i n y} g^*(\log(1+1/y))dy.$$

Inserting these into (4.1.14) we find that for any $0 < \alpha < 1$

$$\begin{aligned}\int_{-\infty}^\infty |\zeta(\alpha+it)|^2 g(t)dt &= \zeta(2\alpha)g^*(0)\\ &+ 2\zeta(2\alpha-1)\mathrm{Re}\{\tilde{g}(2\alpha-1,\alpha)\} - 4\pi\zeta(2\alpha-1)\mathrm{Re}\{g((\alpha-1)i)\}\\ &+ 4\sum_{n=1}^\infty \sigma_{1-2\alpha}(n)\int_0^\infty (y(y+1))^{-\alpha}\cos(2\pi n y)\mathrm{Re}\{g^*(\log(1+1/y))\}dy.\end{aligned}$$

Finally we combine this with (4.1.15), and end the proof of the following explicit formula for the mean square of the zeta-function:

Theorem 4.1 *If the function g satisfies the basic assumption, then we have*

$$\begin{aligned}\mathscr{Z}_1(g) = \int_{-\infty}^\infty \left[\mathrm{Re}\left\{\frac{\Gamma'}{\Gamma}(\tfrac{1}{2}+it)\right\} + 2c_E - \log(2\pi)\right] g(t)dt + 2\pi\mathrm{Re}\{g(\tfrac{1}{2}i)\}\\ + 4\sum_{n=1}^\infty d(n)\int_0^\infty (y(y+1))^{-\frac{1}{2}} g_c(\log(1+1/y))\cos(2\pi n y)dy,\end{aligned} \tag{4.1.16}$$

where

$$g_c(x) = \int_{-\infty}^\infty g(t)\cos(xt)dt.$$

4.2 Non-diagonals

In the subsequent sections we shall develop an argument which extends (4.1.16) to the fourth power moment situation, and demonstrate that the zeta-values are intrinsically related to the eigenvalues of the non-Euclidean Laplacian Δ. Here we shall try to elicit that the emergence of those exotic objects was inevitable in the development of the theory of the zeta-function, and thereby indicate our plan of investigating $\mathscr{Z}_2(g)$.

In number theory we deal with the problem of estimating

$$\sum_n F(n).$$

If F has some tangible algebraic structure then it would be possible to deal with this by means of specific but highly efficient argument. However, most problems are not of such simplicity; and we often have to seek for some transformation to get analytically smoother expressions. In this respect Poisson's sum formula and Dirichlet series expansion are our favorite tools. We then encounter trigonometrical sums like

$$\sum_{m \in I} e(f(n)),$$

where I is an interval and f is a smooth function. Hence methods of estimating trigonometrical sums have been much sought for; and there is a general principle due to H. Weyl. A version of it is attributed to van der Corput, and is built upon the following triviality:

$$\sum_{m \in I} e(f(n)) = \frac{1}{J} \sum_{n} \sum_{j=1}^{J} e(f(n+j))\delta_I(n+j), \tag{4.2.1}$$

where J is arbitrary and δ_I is the characteristic function of I. If some effective inequalities are applied to the right side, then the result does not look trivial any more but becomes a sharp tool. In fact it could yield deep estimates of $\zeta(s)$ that played basic rôles in many problems in number theory, most notably in the theory of prime numbers.

One may perhaps take (4.2.1) for a kind of lifting of a one dimensional sum to a two dimensional sum. The original problem has been transformed into the one of finding interaction among the *non-diagonal* entries. This observation leads us to another triviality: For general double sums we have the decomposition

$$\sum_{m,n} F(m,n) = \left\{ \sum_{m=n} + \sum_{m<n} + \sum_{m>n} \right\} F(m,n), \tag{4.2.2}$$

which is exactly the same as what we did at (4.1.6). Thus, in this context, the formula (4.1.16) could be regarded as a completion of van der Corput's method applied to $\zeta(s)$. In our view the most significant aspect in the proof of (4.1.16) lies in that with the aid of (4.2.2) we can avoid the approximate functional equation for $\zeta(s)$ that was a cause of certain incompleteness in the past results on $\mathscr{Z}_1(g)$. Hence it is appropriate to designate (4.2.2) as the Atkinson dissection, acknowledging its first implementer.

Then one may ask whether the dissection mode in (4.2.2) is always optimal or not. It is certainly not in general. An option to tune it up is to refine the notion of being diagonal. A simple arithmetic way to do this is to replace (4.2.2) by

$$\sum_{m,n} F(m,n) = \Big\{ \sum_{km=ln} + \sum_{km<ln} + \sum_{km>ln} \Big\} F(m,n),$$

where k, l are arbitrary non-zero integers; so the base line is now $y = (l/k)x$ instead of $y = x$. To extract information from all of these dissections we multiply both sides by a weight $w(k,l)$ and sum over all k, l. We then get

$$\sum_{m,n} F(m,n) = \frac{1}{W} \Big\{ \sum_{km=ln} + \sum_{km<ln} + \sum_{km>ln} \Big\} w(k,l) F(m,n), \tag{4.2.3}$$

where the sums on the right side are over the four variables, and W is the sum of all weights. This obviously corresponds to (4.2.1). As before, we may relate it to a lifting of a double sum to a four dimensional sum; and we are led to the following trivial decomposition of generic quadruple sums:

$$\sum_{k,l,m,n} F(k,l,m,n) = \Big\{ \sum_{km=ln} + \sum_{km<ln} + \sum_{km>ln} \Big\} F(k,l,m,n). \tag{4.2.4}$$

Then we take a new viewpoint: We regard quadruple sums as sums over 2×2 integral matrices N. The last identity thus becomes

$$\sum_{N} F(N) = \Big\{ \sum_{|N|=0} + \sum_{|N|>0} + \sum_{|N|<0} \Big\} F(N).$$

Invoking Hecke's representatives mod $SL(2,\mathbb{Z})$ of matrices with given determinants (cf. (3.1.2)), we have further

$$\begin{aligned} \sum_{|N|>0} F(N) &= \sum_{n=1}^{\infty} \sum_{|N|=n} F(N) \\ &= \sum_{n=1}^{\infty} \sum_{ad=n} \sum_{b=1}^{d} \sum_{N \in SL(2,\mathbb{Z})} F\left(N \begin{pmatrix} a & b \\ & d \end{pmatrix} \right). \end{aligned} \tag{4.2.5}$$

Hereby we find an inherent relation lying between the Atkinson dissection and the theory of automorphic forms. If F is sufficiently smooth one may apply the harmonic analysis over $SL(2,\mathbb{Z})$ to the innermost sum in the last identity, and the outer sums are possibly controlled by the theory of

Hecke operators. Or, alternatively, using the ordinary Fourier expansion one may reduce the innermost sum to sums of Kloosterman sums, and thus to the trace formulas (2.4.7) and (2.5.14).

We should, however, be cautious not to be too optimistic about the applicability of our argument. For both (4.2.2) and (4.2.4) take us inevitably to the problem of dealing with non-diagonal elements that may contain subtler arithmetic structure than the diagonal elements. In fact it is often the case that this is equivalent to a problem of the binary additive type

$$\sum_{n\in I} \tau_1(n)\tau_2(n+h) \quad (h \neq 0)$$

with certain arithmetic functions τ_ν; and it is one of the most tantalizing subjects in number theory. The luck that we experienced in deriving (4.1.16) is due to no less than the fact that in that case we have the most trivial situation $\tau_\nu \equiv 1$. Equally felicitous is the case of $\mathscr{Z}_2(g)$, since there we have $\tau_\nu = d$, the divisor function, and the binary additive divisor problem thus arising admits a group structure that can be exploited most suitably with the decomposition (4.2.5).

4.3 Reduction to Kloosterman sums

We are now approaching a pass to the non-Euclidean plane: We shall connect $\mathscr{Z}_2(g)$ with Kloosterman sums and prepare the way towards its spectral decomposition.

To this end let $\mathscr{D}_+$ and $\mathscr{D}_-$ be the domains of $\mathbb{C}^4$ where all four variables have real parts larger than and less than one, respectively. We put, for $(u, v, w, z) \in \mathscr{D}_+$,

$$\mathscr{I}(u, v, w, z; g) = \int_{-\infty}^{\infty} \zeta(u+it)\zeta(v+it)\zeta(w-it)\zeta(z-it)g(t)dt. \qquad (4.3.1)$$

Moving the path upwards appropriately, we see that $\mathscr{I}$ is a meromorphic function over the domain

$$\mathscr{B} = \{ (u, v, w, z) \in \mathbb{C}^4 \ : \ |u|, \ |v|, \ |w|, \ |z| < B \}, \qquad (4.3.2)$$

where B is as in (4.1.2). Then, taking (u, v, w, z) to $\mathscr{D}_- \cap \mathscr{B}$ and shifting the path back to the original, we get the following meromorphic continuation of $\mathscr{I}$ to $D_- \cap \mathscr{B}$:

$$\begin{aligned}\mathscr{I}(u,v,w,z;g) = &\int_{-\infty}^{\infty} \zeta(u+it)\zeta(v+it)\zeta(w-it)\zeta(z-it)g(t)dt \\ &+ 2\pi\zeta(v-u+1)\zeta(u+w-1)\zeta(u+z-1)g((u-1)i) \\ &+ 2\pi\zeta(u-v+1)\zeta(v+w-1)\zeta(v+z-1)g((v-1)i) \\ &+ 2\pi\zeta(z-w+1)\zeta(u+w-1)\zeta(v+w-1)g((1-w)i) \\ &+ 2\pi\zeta(w-z+1)\zeta(u+z-1)\zeta(v+z-1)g((1-z)i),\end{aligned} \tag{4.3.3}$$

which corresponds to (4.1.3).

Lemma 4.2 *The function* $\mathscr{I}$ *is regular at the point*

$$P_{\frac{1}{2}} = (\tfrac{1}{2}, \tfrac{1}{2}, \tfrac{1}{2}, \tfrac{1}{2}); \tag{4.3.4}$$

and we have

$$\mathscr{Z}_2(g) = \mathscr{I}(P_{\frac{1}{2}};g) - 2\pi \mathrm{Re}\,\{(c_E - \log(2\pi))g(\tfrac{1}{2}i) + \tfrac{1}{2}ig'(\tfrac{1}{2}i)\}. \tag{4.3.5}$$

Proof On the right side of (4.3.3) the integral is obviously regular throughout $\mathscr{D}_- \cap \mathscr{B}$. To see the regularity at $P_{\frac{1}{2}}$ of the sum of the other terms we need only to replace the factors $\zeta(v-u+1)$, $\zeta(u-v+1)$, $\zeta(z-w+1)$, and $\zeta(w-z+1)$ by their Laurent expansions; and its value at $P_{\frac{1}{2}}$ is

$$\begin{aligned}&4\pi\zeta(0)^2 c_E\{g(\tfrac{1}{2}i) + g(-\tfrac{1}{2}i)\} \\ &- 2\pi\left[\frac{\partial}{\partial v}\{\zeta(v+w-1)\zeta(v+z-1)g((v-1)i)\}\right. \\ &\left.+ \frac{\partial}{\partial z}\{\zeta(u+z-1)\zeta(v+z-1)g((1-z)i)\}\right]_{P_{\frac{1}{2}}} \\ &= 4\pi\{\zeta(0)^2 c_E - \zeta(0)\zeta'(0)\}\{g(\tfrac{1}{2}i) + g(-\tfrac{1}{2}i)\} \\ &+ 2\pi\zeta(0)^2 i\{g'(\tfrac{1}{2}i) - g'(-\tfrac{1}{2}i)\},\end{aligned}$$

which ends the proof of the lemma.

Because of the identity (4.3.5) we seek for some other way of continuing $\mathscr{I}$ from $\mathscr{D}_+$ to a neighborhood of $P_{\frac{1}{2}}$. For this purpose we note that in $\mathscr{D}_+$

$$\mathscr{I}(u,v,w,z;g) = \sum_{k,l,m,n=1}^{\infty} k^{-u}l^{-v}m^{-w}n^{-z}g^*(\log(mn)/(kl)),$$

where g^* is as before. Hence (4.2.4) comes into play: We have the decomposition, in $\mathscr{D}_+$,

$$\mathscr{J}(u,v,w,z;g) = \mathscr{J}_0(u,v,w,z;g) + \mathscr{J}_1(u,v,w,z;g) + \overline{\mathscr{J}_1(\overline{w},\overline{z},\overline{u},\overline{v};g)}, \tag{4.3.6}$$

where $\mathscr{J}_0$ and $\mathscr{J}_1$ correspond to the parts with $kl = mn$ and $kl < mn$, respectively.

An identity of Ramanujan gives

$$\mathscr{J}_0(u,v,w,z;g) = g^*(0)\zeta(u+w)\zeta(u+z)\zeta(v+w)\zeta(v+z)/\zeta(u+v+w+z), \tag{4.3.7}$$

which is obviously meromorphic over $\mathbb{C}^4$.

As to $\mathscr{J}_1$ we have, in $\mathscr{D}_+$,

$$\mathscr{J}_1(u,v,w,z;g) = \sum_{m,n=1}^{\infty} \frac{\sigma_{u-v}(m)\sigma_{w-z}(m+n)}{m^u(m+n)^w} g^*(\log(1+n/m));$$

and by (4.1.12)

$$\mathscr{J}_1(u,v,w,z;g) = \frac{1}{2\pi i} \sum_{m,n=1}^{\infty} \frac{\sigma_{u-v}(m)\sigma_{w-z}(m+n)}{m^{u+w}} \int_{(2)} \tilde{g}(s,w)(m/n)^s ds. \tag{4.3.8}$$

Here the presence of the factor $\sigma_{w-z}(m+n)$ prevents us going further on this line of reasoning which is an analogue of that of the second section. In effect we need to find a way to separate the variables m and n trapped in $\sigma_{w-z}(m+n)$; and this is the place where the decomposition (4.2.5) turns out to be no abstract nonsense. However, to avoid straying far afield we take a short cut, although it is essentially equivalent to working directly with (4.2.5). To wit, we appeal to the fact that the arithmetic function σ_ξ appears in the Fourier expansion of the Eisenstein series for the full modular group Γ via the identity (1.1.14). Then, to secure absolute convergence, we introduce a subdomain of $\mathscr{D}_+$:

$$\mathscr{D}_{+,0} = \{(u,v,w,z;g) \in \mathscr{D}_+ : \operatorname{Re} z > \operatorname{Re} w + 1 > 3\}.$$

There we have, by (1.1.14) and (4.3.8),

$$\mathscr{J}_1(u,v,w,z;g) = \zeta(z-w+1) \sum_{l=1}^{\infty} l^{w-z-1}$$

$$\times \sum_{\substack{h=1 \\ (h,l)=1}}^{l} \frac{1}{2\pi i} \int_{(2)} D(u+w-s, u-v; e(h/l))\zeta(s, e(h/l))\tilde{g}(s,w) ds, \tag{4.3.9}$$

where the D-function is defined by (3.3.22), and

$$\zeta(s, e(h/l)) = \sum_{n=1}^{\infty} e(nh/l)n^{-s}.$$

We are going to shift the path in (4.3.9) to the right. For this purpose we introduce another domain:

$$\mathscr{E} = \{(u, v, w, z) \in \mathscr{B} : \operatorname{Re}(u + w) < \tfrac{1}{3}B, \\ \operatorname{Re}(v + w) < \tfrac{1}{3}B, \operatorname{Re}(u + v + w + z) > 3B\}. \quad (4.3.10)$$

Obviously we have

$$\mathscr{D}_{+,0} \cap \mathscr{E} \neq \emptyset;$$

and in this joint domain we shall work for a moment. Then we have, by Lemmas 3.7 and 4.1,

$$\sum_{\substack{h=1\\(h,l)=1}}^{l} \frac{1}{2\pi i}\left\{\int_{(2)} - \int_{(B)}\right\} D(u + w - s, u - v; e(h/l))\zeta(s, e(h/l))\tilde{g}(s, w)ds$$

$$= l^{u-v-1}\zeta(v - u + 1)\tilde{g}(u + w - 1, w)\sum_{n=1}^{\infty} c_l(n)n^{1-u-w}$$

$$+ l^{v-u-1}\zeta(u - v + 1)\tilde{g}(v + w - 1, w)\sum_{n=1}^{\infty} c_l(n)n^{1-v-w} \quad (4.3.11)$$

provided $u \neq v$, where c_l is the Ramanujan sum (1.1.15). In the integral over the line $\operatorname{Re} s = B$ we have $\operatorname{Re}(u+w-s) < 0$ and $\operatorname{Re}(v+w-s) < 0$; so we may replace $D(u+w-s, u-v; e(h/l))$ by the absolutely convergent Dirichlet series implied by (3.3.23). This yields, after some rearrangement,

$$\sum_{\substack{h=1\\(h,l)=1}}^{l} \frac{1}{2\pi i}\int_{(B)} D(u + w - s, u - v; e(h/l))\zeta(s, e(h/l))\tilde{g}(s, w)ds$$

$$= 2(2\pi)^{w-z-1}l^{z-w}\sum_{m,n=1}^{\infty} m^{\frac{1}{2}(1-u-v-w-z)}n^{\frac{1}{2}(u+w-v-z-1)}\sigma_{v-u}(n)$$

$$\times\Big\{S(m, n; l)\varphi_+\Big(\frac{4\pi}{l}\sqrt{mn}; u, v, w, z; g\Big)$$

$$+ S(m, -n; l)\varphi_-\Big(\frac{4\pi}{l}\sqrt{mn}; u, v, w, z; g\Big)\Big\}. \quad (4.3.12)$$

Here

$$\varphi_+(x;u,v,w,z;g) = \frac{1}{2\pi i}\cos(\tfrac{1}{2}(u-v)\pi) \times \int_{(B)} \left(\frac{x}{2}\right)^{u+v+w+z-1-2s} \Gamma(s+1-u-w) \times \Gamma(s+1-v-w)\tilde{g}(s,w)ds, \tag{4.3.13}$$

$$\varphi_-(x;u,v,w,z;g) = -\frac{1}{2\pi i}\int_{(B)} \left(\frac{x}{2}\right)^{u+v+w+z-1-2s} \cos(\pi(w+\tfrac{1}{2}(u+v)-s)) \times \Gamma(s+1-u-w)\Gamma(s+1-v-w)\tilde{g}(s,w)ds. \tag{4.3.14}$$

Combining (4.3.9), (4.3.11), and (4.3.12), we obtain

Lemma 4.3 *The function $\mathscr{I}_1(u,v,w,z;g)$ can be continued meromorphically to the domain $\mathscr{E}$, and there we have the decomposition*

$$\mathscr{I}_1(u,v,w,z) = \mathscr{I}_2(u,v,w,z) + \mathscr{I}_3^+(u,v,w,z) + \mathscr{I}_3^-(u,v,w,z). \tag{4.3.15}$$

Here

$$\begin{aligned}\mathscr{I}_2(u,v,w,z;g) = {} & \tilde{g}(u+w-1)\zeta(v+z)\zeta(u+w-1)\zeta(z-w+1) \\ & \times \zeta(v-u+1)/\zeta(v+z-u-w+2), \\ & + \tilde{g}(v+w-1)\zeta(u+z)\zeta(v+w-1)\zeta(z-w+1) \\ & \times \zeta(u-v+1)/\zeta(u+z-v-w+2)\end{aligned} \tag{4.3.16}$$

and

$$\mathscr{I}_3^{\pm}(u,v,w,z;g) = 2(2\pi)^{w-z-1}\zeta(z-w+1) \times \sum_{m,n=1}^{\infty} m^{\frac{1}{2}(1-u-v-w-z)} n^{\frac{1}{2}(u+w-v-z-1)} \sigma_{v-u}(n) K_{\pm}(m,n;u,v,w,z;g), \tag{4.3.17}$$

where

$$K_{\pm}(m,n;u,v,w,z;g) = \sum_{l=1}^{\infty} \frac{1}{l} S(m,\pm n;l)\varphi_{\pm}\left(\frac{4\pi}{l}\sqrt{mn};u,v,w,z;g\right) \tag{4.3.18}$$

with $\varphi_{\pm}$ being defined by (4.3.13) *and* (4.3.14).

Proof The contribution of the right hand side of (4.3.11) to $\mathscr{I}_1$ can be computed with (1.1.14), and is equal to the term $\mathscr{I}_2$. On the other hand it is easy to check that the series in (4.3.17) converges absolutely and uniformly in $\mathscr{E}$, giving the meromorphic continuation of $\mathscr{J}_1$. We remark that we need here only trivial bounds for Kloosterman sums. This ends the proof of the lemma.

This finishes the reduction of the problem $\mathscr{Z}_2(g)$ to sums of Kloosterman sums. What we are going to undertake next is to continue $\mathscr{J}_1$ to the domain $\mathscr{B}$. It should be stressed that this corresponds exactly to the situation we had at (4.1.13), although there the relevant analytic continuation of $\mathscr{J}_1$ was almost a triviality. Our new problem is certainly far deeper, and will be resolved only after decomposing $K_\pm(m,n;u,v,w,z;g)$ spectrally.

4.4 Spectral expansion

We are going to complete the double summation involved in the formula (4.3.17). This will be achieved by separating the summation variables, which amounts to an application of the trace formulas (2.4.7) and (2.5.14) to $K_\pm(m,n;u,v,w,z;g)$. The philosophy behind trace formulas or spectral expansions in general is to exploit hidden orthogonalities in a given problem and gain better handling of variables by separating them via identities of the Parseval type. This is exactly the case with our present situation, and we shall see that the separation of the variables m,n in $K_\pm(m,n;u,v,w,z;g)$ thus performed enables us to reduce the problem further to the analytical properties of automorphic L-functions as we indicated already at the beginning of the previous chapter. That these functions are entire and of controllable size is particularly important in developing our argument. For it entails an expansion of the domain of the existence of $\mathscr{J}_3^\pm(u,v,w,z;g)$ to the one that is symmetric in the four variables and contains the critical point $P_{\frac{1}{2}}$. Here it should be stressed that the symmetry is essential for our purpose, since we have the decomposition (4.3.6) which prevents us practicing discrimination among the variables.

Before entering into discussion we recollect, for the purpose of convenience, what we have done on the choice of the orthonormal system of the cusp-forms over the full modular group. Thus the elements in the system $\{\psi_j : j \geq 1\}$ introduced at (1.1.43) are assumed to be such that their Maass–Fourier coefficients $\rho_j(n)$ satisfy (3.1.16), i.e.,

$$\rho_j(n) = \rho_j(1)t_j(n), \quad \rho(-n) = \varepsilon_j\rho_j(n), \tag{4.4.1}$$

where the Hecke eigenvalues $t_j(n)$ and the parity sign ε_j are defined by (3.1.13) and (3.1.15), respectively. Also the elements in the system $\{\psi_{j,k} : 1 \le j \le \vartheta(k)\}$ introduced at (2.2.2)–(2.2.3) have the Fourier coefficients $\rho_{j,k}(n)$ satisfying (3.1.21), i.e.,

$$\rho_{j,k}(n) = \rho_{j,k}(1)t_{j,k}(n), \tag{4.4.2}$$

where the Hecke eigenvalues $t_{j,k}(n)$ are defined by (3.1.20). Further, we put

$$\alpha_j = (\cosh \pi\kappa_j)^{-1}|\rho_j(1)|^2, \quad \alpha_{j,k} = a(k)|\rho_{j,k}(1)|^2 \tag{4.4.3}$$

with $a(k)$ as in (2.2.9), i.e., $a(k) = 2^{1-4k}\pi^{-2k-1}\Gamma(2k)$. The symbol α_j has already been introduced at (3.3.1). We have, from (3.1.18) and (3.1.22),

$$t_j(n),\ t_{j,k}(n) \ll n^{\frac{1}{4}+\delta} \tag{4.4.4}$$

with the implied constant depending only on δ, an arbitrary small positive constant. Also we have, from (3.3.2) and (2.2.10),

$$\sum_{K\le\kappa_j<2K} \alpha_j \ll K^2, \qquad \sum_{j=1}^{\vartheta(k)} \alpha_{j,k} \ll k, \tag{4.4.5}$$

respectively, where the implied constants are absolute. Further, we stress that the basic assumption is imposed upon the weight function g; thus the parameter A will appear in this context.

Now, we begin with $K_+(m,n;u,v,w,z;g)$, to which we apply (2.4.7). We shall keep the condition $(u,v,w,z) \in \mathscr{E}$ for a while, where $\mathscr{E}$ is as in (4.3.10). First we have to check whether $\varphi_+(x;u,v,w,z;g)$ defined by (4.3.13) satisfies the condition (2.4.6). The three-time differentiability and the decay condition at the point $x = 0$ are trivially fulfilled. The decay condition at positive infinity is also satisfied as can be seen by shifting the path $\mathrm{Re}\, s = B$ in (4.3.13) to $\mathrm{Re}\, s = \frac{1}{3}A$ while appealing to Lemma 4.1. Hence we have, with the conventions (4.4.1)–(4.4.3),

$$\begin{aligned} K_+(m,n;u,v,w,z;g) &= \sum_{j=1}^{\infty} \alpha_j t_j(m)t_j(n)(\varphi_+)^+(\kappa_j;u,v,w,z;g) \\ &+ \frac{1}{\pi}\int_{-\infty}^{\infty} \frac{\sigma_{2ir}(m)\sigma_{2ir}(n)}{(mn)^{ir}|\zeta(1+2ir)|^2}(\varphi_+)^+(r;u,v,w,z;g)dr \\ &+ 2\sum_{k=1}^{\infty}\sum_{j=1}^{\vartheta(k)} \alpha_{j,k}t_{j,k}(m)t_{j,k}(n)(\varphi_+)^+((\tfrac{1}{2}-k)i;u,v,w,z;g), \end{aligned} \tag{4.4.6}$$

where

$$(\varphi_+)^+(r;u,v,w,z;g) = \frac{\pi i}{2\sinh(\pi r)}\int_0^\infty (J_{2ir}(x)-J_{-2ir}(x))\varphi_+(x;u,v,w,z;g)\frac{dx}{x}.$$

To transform the integral we consider, in view of (4.3.13),

$$\int_0^\infty J_{2ir}(x)\int_{(B)}\Gamma(s+1-u-w)\Gamma(s+1-v-w) \times \tilde{g}(s,w)\left(\frac{x}{2}\right)^{u+v+w+z-2-2s} dsdx, \tag{4.4.7}$$

where $\tilde{g}$ is defined by (4.1.7). The double integral is regular in the domain

$$\left\{(u,v,w,z): \begin{array}{c}\mathrm{Re}\,(u+w)<B+1,\ \mathrm{Re}\,(v+w)<B+1,\\ 2B+1<\mathrm{Re}\,(u+v+w+z)<\frac{2}{3}A+\frac{3}{2}\end{array}\right\}, \tag{4.4.8}$$

which contains $\mathscr{E}$. To see this we divide (4.4.7) into two parts according as $0\le x<1$ and $x\ge 1$, and note that

$$J_{2ir}(x) \ll \begin{cases} 1 & \text{as } x\to +0,\\ x^{-\frac{1}{2}} & \text{as } x\to+\infty. \end{cases} \tag{4.4.9}$$

The first part is clearly regular in (4.4.8); and in the second part we move the path in the s-integral to $\mathrm{Re}\,s=\frac{1}{3}A$, which gives the assertion. We then consider the subdomain of (4.4.8) where we have $2B+1<\mathrm{Re}\,(u+v+w+z)<2B+\frac{3}{2}$. There (4.4.7) is absolutely convergent; and it is equal to

$$\int_{(B)}\frac{\Gamma(\frac{1}{2}(u+v+w+z-1)-s+ir)}{\Gamma(\frac{1}{2}(3-u-v-w-z)+s+ir)}\Gamma(s+1-u-w) \times\Gamma(s+1-v-w)\tilde{g}(s,w)ds. \tag{4.4.10}$$

Here we have used the fact that

$$\int_0^\infty J_\nu(x)\left(\frac{x}{2}\right)^{-\mu}dx = \frac{\Gamma(\frac{1}{2}(\nu+1-\mu))}{\Gamma(\frac{1}{2}(\nu+1+\mu))} \tag{4.4.11}$$

provided $\mathrm{Re}\,\mu>\frac{1}{2}$, $\mathrm{Re}\,(\nu-\mu)>-1$, which is an inversion of the relation (2.2.17). But the integral (4.4.10) is obviously regular in (4.4.8); and by analytic continuation (4.4.7) is equal to (4.4.10) throughout the domain

(4.4.8). Hence, after a rearrangement, we have in $\mathscr{E}$

$$\begin{aligned}(\varphi_+)^+(r;\, u, v, w, z; g) &= \frac{1}{4\pi i}\cos(\tfrac{1}{2}\pi(u-v))\int_{(B)}\sin(\tfrac{1}{2}\pi(u+v+w+z-2s)) \\ &\times \Gamma(\tfrac{1}{2}(u+v+w+z-1)+ir-s)\Gamma(\tfrac{1}{2}(u+v+w+z-1)+ir-s) \\ &\times \Gamma(s+1-u-w)\Gamma(s+1-v-w)\tilde{g}(s,w)ds. \qquad (4.4.12)\end{aligned}$$

As to $(\varphi_+)^+((\frac{1}{2}-k)i; u, v, w, z; g)$, we note that $J_{1-2k}(x) = -J_{2k-1}(x)$ for any integer $k \geq 1$. Then we can show just as above that for any integer $k \geq 1$ and $(u, v, w, z) \in \mathscr{E}$

$$\begin{aligned}(\varphi_+)^+((\tfrac{1}{2}-k)i; u, v, w, z; g) &= \frac{(-1)^k}{4i}\cos(\tfrac{1}{2}\pi(u-v)) \\ \times \int_{(B)} &\frac{\Gamma(k+\frac{1}{2}(u+v+w+z-2)-s)}{\Gamma(k+\frac{1}{2}(2-u-v-w-z)+s)} \\ &\times \Gamma(s+1-u-w)\Gamma(s+1-v-w)\tilde{g}(s,w)ds. \qquad (4.4.13)\end{aligned}$$

Next we consider $K_-(m,n;u,v,w,z;g)$ briefly. As in the case of $\varphi_+(x;u,v,w,z;g)$ we see readily that $\varphi_-(x;u,v,w,z;g)$ satisfies (2.4.6); yet there is an essential difference between them, that is, this time the bound (4.1.10) is indispensable. The trace formula (2.5.14) with the convention (4.4.1) yields

$$\begin{aligned}K_-(m,n;u,v,w,z;g) &= \sum_{j=1}^{\infty}\varepsilon_j\alpha_j t_j(m)t_j(n)(\varphi_-)^-(\kappa_j;u,v,w,z;g) \\ &+ \frac{1}{\pi}\int_{-\infty}^{\infty}\frac{\sigma_{2ir}(m)\sigma_{2ir}(n)}{(mn)^{ir}|\zeta(1+2ir)|^2}(\varphi_-)^-(r;u,v,w,z;g)dr, \qquad (4.4.14)\end{aligned}$$

where $(u,v,w,z) \in \mathscr{E}$ and

$$(\varphi_-)^-(r;u,v,w,z;g) = 2\cosh(\pi r)\int_0^{\infty}\varphi_-(r;u,v,w,z;g)K_{2ir}(x)\frac{dx}{x}.$$

Inserting (4.3.14) into the last integral we get an absolutely convergent double integral, since we have (4.1.10) and

$$K_{2ir}(x) \ll \begin{cases} |\log x| & \text{as } x \to +0, \\ e^{-x} & \text{as } x \to +\infty \end{cases}$$

provided r is real, which follows readily from (1.1.17)–(1.1.18). Exchanging the order of integrals, we have

$$(\varphi_-)^{-}(r;u,v,w,z;g) = -\frac{1}{4\pi i}\cosh(\pi r)\int_{(B)}\cos(\pi(w+\tfrac{1}{2}(u+v)-s))$$
$$\times\Gamma(s+1-u-w)\Gamma(s+1-v-w)\tilde{g}(s,w)$$
$$\times\int_0^\infty K_{2ir}(x)\left(\frac{x}{2}\right)^{u+v+w+z-2-2s}dxds;$$

and thus, by (2.6.12),

$$(\varphi_-)^{-}(r;u,v,w,z;g)$$
$$= -\frac{1}{4\pi i}\cosh(\pi r)\int_{(B)}\cos(\pi(w+\tfrac{1}{2}(u+v)-s))$$
$$\times\Gamma(\tfrac{1}{2}(u+v+w+z-1)+ir-s)\Gamma(\tfrac{1}{2}(u+v+w+z-1)-ir-s)$$
$$\times\Gamma(s+1-u-w)\Gamma(s+1-v-w)\tilde{g}(s,w)ds. \tag{4.4.15}$$

Now, in order to facilitate later discussion, we introduce three functions $\Phi_\pm$ and Ξ of five complex variables:

$$\Phi_+(\xi;u,v,w,z;g)$$
$$= -i(2\pi)^{w-z-2}\cos(\tfrac{1}{2}\pi(u-v))\int_{-i\infty}^{i\infty}\sin(\tfrac{1}{2}\pi(u+v+w+z-2s))$$
$$\times\Gamma(\tfrac{1}{2}(u+v+w+z-1)+\xi-s)\Gamma(\tfrac{1}{2}(u+v+w+z-1)+\xi-s)$$
$$\times\Gamma(s+1-u-w)\Gamma(s+1-v-w)\tilde{g}(s,w)ds; \tag{4.4.16}$$

$$\Phi_-(\xi;u,v,w,z;g)$$
$$= i(2\pi)^{w-z-2}\cos(\pi\xi)\int_{-i\infty}^{i\infty}\cos(\pi(w+\tfrac{1}{2}(u+v)-s))$$
$$\times\Gamma(\tfrac{1}{2}(u+v+w+z-1)+\xi-s)\Gamma(\tfrac{1}{2}(u+v+w+z-1)-\xi-s)$$
$$\times\Gamma(s+1-u-w)\Gamma(s+1-v-w)\tilde{g}(s,w)ds; \tag{4.4.17}$$

$$\Xi(\xi;u,v,w,z;g)$$
$$= \frac{1}{2\pi i}\int_{-\infty i}^{\infty i}\frac{\Gamma(\xi+\frac{1}{2}(u+v+w+z-1)-s)}{\Gamma(\xi+\frac{1}{2}(3-u-v-w-z)+s)}$$
$$\times\Gamma(s+1-u-w)\Gamma(s+1-v-w)\tilde{g}(s,w)ds. \tag{4.4.18}$$

Here the path in (4.4.16) is such that the poles of the first two gamma-factors and those of the other three factors in the integrand are separated to the right and the left, respectively, by the path, and ξ, u, v, w, z are assumed to be such that the path can be drawn. The path in (4.4.17) is chosen in just the same way. On the other hand the path in (4.4.18) separates the poles of $\Gamma(\xi + \frac{1}{2}(u + v + w + z - 1) - s)$ and those of $\Gamma(s + 1 - u - w)\Gamma(s + 1 - v - w)\tilde{g}(s, w)$ to the left and the right of the path, respectively.

Lemma 4.4 *We have*

$$\Phi_+(\xi; u, v, w, z; g) = -\frac{(2\pi)^{w-z} \cos\left(\frac{1}{2}\pi(u - v)\right)}{4\sin(\pi\xi)} \times \{\Xi(\xi; u, v, w, z; g) - \Xi(-\xi; u, v, w, z; g)\}; \tag{4.4.19}$$

$$\Phi_-(\xi; u, v, w, z; g) = \frac{(2\pi)^{w-z}}{4\sin(\pi\xi)}\{\sin(\pi(\tfrac{1}{2}(z - w) + \xi))\Xi(\xi; u, v, w, z; g) - \sin(\pi(\tfrac{1}{2}(z - w) - \xi))\Xi(-\xi; u, v, w, z; g)\}, \tag{4.4.20}$$

provided the left sides are well-defined. Also we have, for real r and $(u, v, w, z) \in \mathscr{E}$,

$$(\varphi_+)^+(r; u, v, w, z; g) = \tfrac{1}{2}(2\pi)^{1-w+z}\Phi_+(ir; u, v, w, z; g), \tag{4.4.21}$$

$$(\varphi_-)^-(r; u, v, w, z; g) = \tfrac{1}{2}(2\pi)^{1-w+z}\Phi_-(ir; u, v, w, z; g); \tag{4.4.22}$$

and, for integral $k \geq 1$ and $(u, v, w, z) \in \mathscr{E}$,

$$(\varphi_+)^+(i(\tfrac{1}{2} - k); u, v, w, z; g) = \tfrac{1}{2}(-1)^k\pi\cos(\tfrac{1}{2}\pi(u - v))\Xi(k - \tfrac{1}{2}; u, v, w, z; g). \tag{4.4.23}$$

Proof In fact, if $\operatorname{Re}\xi = 0$ and $(u, v, w, z) \in \mathscr{E}$ then we can use the path $\operatorname{Re} s = B$ in (4.4.16) and (4.4.17); thus (4.4.21) and (4.4.22) follow from (4.4.12) and (4.4.15), respectively, while the formula (4.4.23) becomes the same as (4.4.13) with this choice of the path in (4.4.18). As to the first two identities we shall prove only the second; for the other is easier. Thus, if ξ, u, v, w, z are such that the left side of (4.4.20) is well-defined then both $\Xi(\pm\xi; u, v, w, z; g)$ are well-defined, for the path in (4.4.17) can be used in (4.4.18) regardless of ξ being replaced by $-\xi$ or not. With such a choice of the path we see that the expression in the braces on the right side of (4.4.20) is equal to

$$\frac{1}{2\pi i}\int_{-\infty i}^{\infty i}\Big\{\sin(\pi(\tfrac{1}{2}(z-w)+\xi))\frac{\Gamma(\xi+\frac{1}{2}(u+v+w+z-1)-s)}{\Gamma(\xi+\frac{1}{2}(3-u-v-w-z)+s)}$$
$$-\sin(\pi(\tfrac{1}{2}(z-w)-\xi))\frac{\Gamma(-\xi+\frac{1}{2}(u+v+w+z-1)-s)}{\Gamma(-\xi+\frac{1}{2}(3-u-v-w-z)+s)}\Big\}$$
$$\times\Gamma(s+1-u-w)\Gamma(s+1-v-w)\tilde{g}(s,w)ds.$$

The use of the relation $\Gamma(s)\Gamma(1-s)=\pi/\sin(\pi s)$ transforms the expression in these new braces into

$$\pi^{-1}\Gamma(\tfrac{1}{2}(u+v+w+z-1)+\xi-s)\Gamma(\tfrac{1}{2}(u+v+w+z-1)-\xi-s)$$
$$\times\Big\{\sin(\pi(\tfrac{1}{2}(z-w)+\xi))\sin(\pi(\tfrac{1}{2}(u+v+w+z-1)-\xi-s))$$
$$-\sin(\pi(\tfrac{1}{2}(z-w)-\xi))\sin(\pi(\tfrac{1}{2}(u+v+w+z-1)+\xi-s))\Big\};$$

and the rest of the proof is a simple manipulation of trigonometrics.

4.5 Crystallization

Here we shall observe a beautiful process catalyzed by Hecke operators. We are going to insert the spectral expansions (4.4.6) and (4.4.14) into (4.3.17) and exchange the order of sums and integrals.

The absolute convergence that we have to check is obvious as far as the double summation over the variables m,n is concerned, since we have (4.4.4) and $(u,v,w,z)\in\mathscr{E}$. Thus the issue is reduced to bounding $(\varphi_\pm)^\pm$; and Lemma 4.4 renders it in terms of the function Ξ. We then show that we have, uniformly for any fixed compact subset of $\mathscr{E}$,

$$\Xi(ir;u,v,w,z;g)\ll|r|^{-\frac{1}{4}A},\quad \Xi(k-\tfrac{1}{2};u,v,w,z;g)\ll k^{-\frac{1}{4}A} \tag{4.5.1}$$

as real r and positive integral k tend to infinity. To confirm the first assertion we move the path in (4.4.18), which can be assumed to be $\mathrm{Re}\,s=B$ in the present case, to the one consisting of the straight lines connecting the points $B-i\infty$, $B-\frac{1}{2}|r|i$, $\frac{1}{3}A-\frac{1}{2}|r|i$, $\frac{1}{3}A+\frac{1}{2}|r|i$, $B+\frac{1}{2}|r|i$, and $B+i\infty$, where A is as above. If $|r|$ is larger than a constant determined solely by the compact subset of $\mathscr{E}$ under consideration, on this change of the path we do not encounter any singularities. Then on the new path we apply Stirling's formula and (4.1.10) to the integrand, getting the inequality immediately. As to the second assertion we need only to move the path to $\mathrm{Re}\, s = \frac{1}{4}A$, provided k is larger than A, say.

Hence, on noting (4.4.5), we are now able to insert (4.4.6) and (4.4.14) into (4.3.17) and change the order of sums and integrals freely as long as we work inside $\mathscr{E}$. Performing the summation over the variable n, we appreciate the multiplicativity of the Hecke operators reflected in the identities (3.2.7) and (3.2.24). They transform the expressions, which would otherwise remain crude and chaotic, into a harmonious entity. Equally to be appreciated is that, summing over the other variable m, the whole contribution of the terms off the diagonal is absorbed into a single Hecke L-function at each eigen-space. Thus we gain a great ease in controlling the non-diagonals; in fact we can forget about them. Although less significant, the fact that the factor $\zeta(z-w+1)$ on the right side of (4.3.17) is eliminated by virtue of (3.2.7) and (3.2.24) should also be observed properly, for this factor is singular at the point $P_{\frac{1}{2}}$.

Before stating our new expressions for $\mathscr{I}_3^{\pm}$ we put

$$\begin{aligned} S(\xi;u,v,w,z) = {} & \zeta(\tfrac{1}{2}(u+v+w+z-1)+\xi)\zeta(\tfrac{1}{2}(u+v+w+z-1)-\xi) \\ & \times\zeta(\tfrac{1}{2}(u+z-v-w+1)+\xi)\zeta(\tfrac{1}{2}(u+z-v-w+1)-\xi) \\ & \times\zeta(\tfrac{1}{2}(v+z-u-w+1)+\xi)\zeta(\tfrac{1}{2}(v+z-u-w+1)-\xi). \end{aligned} \tag{4.5.2}$$

Then we have obtained

Lemma 4.5 *With the conventions* (4.4.1)–(4.4.3) *we have, in the domain* $\mathscr{E}$,

$$\begin{aligned} \mathscr{I}_3^{+}(u,v,w,z;g) = {} & \mathscr{I}_{3,c}^{+}(u,v,w,z;g) \\ & + \mathscr{I}_{3,d}^{+}(u,v,w,z;g) + \mathscr{I}_{3,h}^{+}(u,v,w,z;g), \end{aligned} \tag{4.5.3}$$

where

$$\mathscr{I}_{3,c}^{+}(u,v,w,z;g) = \frac{1}{i\pi}\int_{(0)} \frac{S(\xi;u,v,w,z)}{\zeta(1+2\xi)\zeta(1-2\xi)}\Phi_{+}(\xi;u,v,w,z;g)d\xi, \tag{4.5.4}$$

$$\begin{aligned} & \mathscr{I}_{3,d}^{+}(u,v,w,z;g) \\ & = \sum_{j=1}^{\infty}\alpha_j H_j(\tfrac{1}{2}(u+v+w+z-1))H_j(\tfrac{1}{2}(u+z-v-w+1)) \\ & \times H_j(\tfrac{1}{2}(v+z-u-w+1))\Phi_{+}(i\kappa_j;u,v,w,z;g), \end{aligned} \tag{4.5.5}$$

$$\mathscr{J}_{3,h}^{+}(u,v,w,z;g) = (2\pi)^{w-z}\cos(\tfrac{1}{2}(u-v))$$
$$\times \sum_{k=6}^{\infty}\sum_{j=1}^{\vartheta(k)}(-1)^k \alpha_{j,k} H_{j,k}(\tfrac{1}{2}(u+v+w+z-1))H_{j,k}(\tfrac{1}{2}(u+z-v-w+1))$$
$$\times H_{j,k}(\tfrac{1}{2}(v+z-u-w+1))\Xi(k-\tfrac{1}{2};u,v,w,z;g). \tag{4.5.6}$$

Lemma 4.6 *With the conventions* (4.4.1) *and* (4.4.3) *we have, in the domain* $\mathscr{E}$,

$$\mathscr{J}_3^{-}(u,v,w,z;g) = \mathscr{J}_{3,c}^{-}(u,v,w,z;g) + \mathscr{J}_{3,d}^{-}(u,v,w,z;g), \tag{4.5.7}$$

where

$$\mathscr{J}_{3,c}^{-}(u,v,w,z;g) = \frac{1}{i\pi}\int_{(0)} \frac{S(\xi;u,v,w,z)}{\zeta(1+2\xi)\zeta(1-2\xi)}\Phi_-(\xi;u,v,w,z;g)d\xi, \tag{4.5.8}$$

$$\mathscr{J}_{3,d}^{-}(u,v,w,z;g)$$
$$= \sum_{j=1}^{\infty}\varepsilon_j\alpha_j H_j(\tfrac{1}{2}(u+v+w+z-1))H_j(\tfrac{1}{2}(u+z-v-w+1))$$
$$\times H_j(\tfrac{1}{2}(v+z-u-w+1))\Phi_-(i\kappa_j;u,v,w,z;g). \tag{4.5.9}$$

4.6 Analytic continuation

The aim of this section is to show that the spectral expansions of $\mathscr{J}_3^{\pm}$ obtained in the preceding section can be continued to the domain $\mathscr{B}$ introduced at (4.3.2), whereby we shall finish our meromorphic continuation of $\mathscr{J}_1$. The domain $\mathscr{B}$ is obviously symmetric and wide enough to have a joint domain with $\mathscr{D}_+$ where the decomposition (4.3.6) was introduced. Hence (4.3.6) should hold throughout $\mathscr{B}$, and we shall obtain a spectral decomposition of $\mathscr{Z}_2(g)$.

By virtue of Lemma 4.4 our problem is equivalent to studying the analytical properties of the function Ξ ; and we are going to show that it is meromorphic in a fairly wide domain in $\mathbb{C}^5$:

Lemma 4.7 *The function* $\Xi(\xi;u,v,w,z;g)$ *is meromorphic in the domain*

$$\widetilde{\mathscr{B}} = \{\xi\ :\mathrm{Re}\,\xi > -\tfrac{1}{8}A\} \times \mathscr{B} \tag{4.6.1}$$

and regular in $\widetilde{\mathscr{B}} \setminus \mathcal{N}$, *where*

$$\mathcal{N} = \big\{(\xi, u, v, w, z) : \text{at least one of } \xi + \tfrac{1}{2}(u+v+w+z-1),\ \xi + \tfrac{1}{2}(u+z-v-w+1) \text{ and } \xi + \tfrac{1}{2}(v+z-u-w+1) \text{ is equal to a non-positive integer}\big\}. \tag{4.6.2}$$

Moreover, if $|\xi|$ *tends to infinity in any fixed vertical or horizontal strips while satisfying* $\operatorname{Re}\xi > -\frac{1}{8}A$, *then we have*

$$\Xi(\xi; u, v, w, z; g) \ll |\xi|^{-\frac{1}{4}A} \tag{4.6.3}$$

uniformly in $\mathscr{B}$.

Proof Intuitively this fact can be inferred from the definition (4.4.18) by just deforming the path appropriately. However, the topological situation involved here is somewhat complicated; thus we employ an explicit argument to avoid any ambiguities.

To this end we put $\mathcal{N}^* = \big[\{\xi : \operatorname{Re}\xi > -\frac{1}{4}A\} \times \mathscr{B}\big] \setminus \mathcal{N}$, which is open and arcwise connected. This can be proved easily by connecting two points of $\mathcal{N}^*$ by a straight line with possible indents. If $(\xi, u, v, w, z) \in \mathcal{N}^*$, then we can obviously draw a path which is needed in (4.4.18) and contained in the vertical strip $|\operatorname{Re}s| \le \frac{1}{3}A$; note that here we have to invoke Lemma 4.1. Thus Ξ is well-defined at all points of $\mathcal{N}^*$. Then, by a routine argument we can show that Ξ is regular and single-valued over $\mathcal{N}^*$. That is, starting at a point of $\mathcal{N}^*$, the function Ξ can be continued analytically to any point of $\mathcal{N}^*$, and the result is always given by the representation (4.4.18) with a suitable choice of the path. Having this, we confine (ξ, u, v, w, z) to the domain defined by the condition

$$\operatorname{Re}\xi > 0,\ (u, v, w, z) \in \mathscr{E}. \tag{4.6.4}$$

Then we may use the line $\operatorname{Re}s = B$ as the path in (4.4.18); the above domain is obviously a subset of $\mathcal{N}^*$. On setting this, we insert (4.1.7) into (4.4.18). The resulting double integral is absolutely convergent, for we have $g^*(\log(1+y)) \ll (1+y)^{-A}$. Interchanging the order of integration, we have

$$\Xi(\xi; u, v, w, z; g) = \int_0^\infty y^{-1}(1+y)^{-w} g^*(\log(1+y)) G(y; \xi; u, v, w, z)\,dy, \tag{4.6.5}$$

where

$$G(y;\xi;u,v,w,z) = \frac{1}{2\pi i}\int_{(B)} \frac{\Gamma(\frac{1}{2}(u+v+w+z-1)+\xi-s)}{\Gamma(\frac{1}{2}(3-u-v-w-z)+\xi+s)} \times \Gamma(s+1-u-w)\Gamma(s+1-v-w)y^s ds, \quad (4.6.6)$$

or rather, using the hypergeometric function F,

$$G(y;\xi;u,v,w,z) = \Gamma(\alpha)\Gamma(\beta)\Gamma(\gamma)^{-1}F(\alpha,\beta;\gamma;-y)y^{\xi+\frac{1}{2}(u+v+w+z-1)} \quad (4.6.7)$$

with

$$\alpha = \xi + \tfrac{1}{2}(u+z-v-w+1), \quad \beta = \xi + \tfrac{1}{2}(v+z-u-w+1), \quad \gamma = 1+2\xi. \quad (4.6.8)$$

We then invoke the integral representation (3.3.40) of the hypergeometric function. Thus, if $0 \le y < 1$, we have, instead of (4.6.6),

$$G(y;\xi;u,v,w,z) = \frac{\Gamma(\xi+\frac{1}{2}(v+z-u-w+1))}{\Gamma(\xi+\frac{1}{2}(v+w-u-z+1))}y^{\xi+\frac{1}{2}(u+v+w+z-1)} \times \int_0^1 x^{\xi+\frac{1}{2}(u+z-v-w-1)}(1-x)^{\xi+\frac{1}{2}(v+w-u-z-1)} \times (1+xy)^{-\xi+\frac{1}{2}(u+w-v-z-1)}dx. \quad (4.6.9)$$

Here both sides are obviously regular for $\mathrm{Re}\, y > 0$; hence this holds, in particular, for all $y \ge 0$ by analytic continuation. We next transform (4.6.5) by applying (3.4.15) with $P = [\frac{1}{2}A]$ to the last factor of the integrand of (4.6.9). Then we get a new expression for $G(y;\xi;u,v,w,z)$, which consists of a sum with P terms and an explicit remainder term. This sum is exactly the sum of the residues of the first P poles of the integrand in (4.6.6) which are on the right of the path. Inserting this result on G into (4.6.5), we find that

$$\Xi(\xi;u,v,w,z) = \sum_{j=0}^{P-1}(-1)^j \frac{\tilde{g}(j+\xi+\frac{1}{2}(u+v+w+z-1),w)}{\Gamma(j+1)\Gamma(j+1+2\xi)} \times \Gamma(j+\xi+\frac{1}{2}(u+z-v-w+1))\Gamma(j+\xi+\frac{1}{2}(v+z-u-w+1)) + (-1)^P \frac{\Gamma(P+\xi+\frac{1}{2}(v+z-u-w+1))}{\Gamma(P)\Gamma(\xi+\frac{1}{2}(v+w-u-z+1))} \times \int_0^1 (1-x)^{\xi+\frac{1}{2}(v+w-u-z-1)} R_P(x;\xi;u,v,w,z)dx \quad (4.6.10)$$

with

$$R_P(x;\xi;u,v,w,z)$$
$$= x^{P+\xi+\frac{1}{2}(u+z-v-w-1)} \int_0^1 \int_0^\infty (1-\theta)^{P-1} y^{P+\xi+\frac{1}{2}(u+v+w+z-3)}$$
$$\times (1+xy\theta)^{-P-\xi+\frac{1}{2}(u+w-v-z-1)}(1+y)^{-w} g^*(\log(1+y))dyd\theta.$$

So far we have assumed (4.6.4); but we may now drop it. Indeed (4.6.10) yields a meromorphic continuation of Ξ to the domain

$$\left\{(\xi,u,v,w,z) : |\mathrm{Re}\,\xi| < \tfrac{1}{8}A,\ \max\{|\mathrm{Re}\,u|,\ |\mathrm{Re}\,v|,\ |\mathrm{Re}\,w|,\ |\mathrm{Re}\,z|\} < \tfrac{1}{24}A\right\}. \tag{4.6.11}$$

The meromorphy of the first P terms in (4.6.10) is obvious. In the last term of (4.6.10) we perform partial integration $[\frac{1}{4}A]$ times. We see readily that the resulting integral converges absolutely and uniformly in the domain (4.6.11), whence Ξ is meromorphic in (4.6.11). On the other hand, if $\mathrm{Re}\,\xi \geq \frac{1}{8}A$ then we can choose the line $\mathrm{Re}\,s = B$ as the path in (4.4.18) whenever $(u,v,w,z) \in \mathscr{B}$. Hence we have finished the proof of the first assertion in the lemma. The decay property (4.6.3) can be proved in much the same way as (4.5.1); so we may omit the details.

We note that in the above we have also proved

Lemma 4.8 *If (ξ,u,v,w,z) is such that the path in* (4.4.18) *can be drawn in a vertical strip contained in the half plane* $\mathrm{Re}\,s > 0$, *then we have*

$$\Xi(\xi;u,v,w,z;g) = \frac{\Gamma(\alpha)\Gamma(\beta)}{\Gamma(\gamma)}$$
$$\times \int_0^\infty \frac{y^{\xi+\frac{1}{2}(u+v+w+z-3)}}{(1+y)^w} g^*(\log(1+y))F(\alpha,\beta;\gamma;-y)dy, \tag{4.6.12}$$

where F is the hypergeometric function and α, β, γ are as in (4.6.8).

Now let us assume that $(u,v,w,z) \in \mathscr{B}$, and examine the consequence of Lemma 4.7 for $\mathscr{I}^+_{3,d}$ defined by (4.5.5): If $\kappa_j \geq 3B$ then $\Xi(\pm i\kappa_j;u,v,w,z;g)$ is regular and $O(\kappa_j^{-\frac{1}{4}A})$ uniformly in $\mathscr{B}$. Hence by (4.4.19) and (4.4.21) the factor $(\varphi_+)^+(\kappa_j;u,v,w,z;g)$ is regular and of exponential decay with respect to κ_j uniformly in $\mathscr{B}$. Since we have (3.2.5) and (4.4.5) this implies immediately that $\mathscr{I}^+_{3,d}$ exists as a meromorphic function inside $\mathscr{B}$. The same can be said about $\mathscr{I}^+_{3,h}$ defined by (4.5.6); we only need to invoke

(3.2.22), (4.4.5), and (4.4.23). As to $\mathscr{I}_{3,d}^{-}$ defined at (4.5.9) we need (4.6.3) essentially; thus via (4.4.20) and (4.4.22) it yields $(\varphi_-)^-(\kappa_j;u,v,w,z;g) = O(\kappa_j^{-\frac{1}{4}A})$ uniformly in $\mathscr{B}$ provided $\kappa_j \geq 3B$. Hence $\mathscr{I}_{3,d}^{-}$ is meromorphic inside $\mathscr{B}$.

Thus it remains for us to consider $\mathscr{I}_{3,c}^{\pm}$ defined at (4.5.4) and (4.5.8), respectively. To this end we assume first that (u,v,w,z) is in $\mathscr{E}$; and put

$$\mathscr{I}_{3,c}(u,v,w,z;g) = \mathscr{I}_{3,c}^{+}(u,v,w,z;g) + \mathscr{I}_{3,c}^{-}(u,v,w,z;g). \tag{4.6.13}$$

We have, by (4.4.19) and (4.4.20),

$$\mathscr{I}_{3,c}(u,v,w,z;g) = i(2\pi)^{w-z-1}\int_{(0)} \frac{S(\xi;u,v,w,z)}{\sin(\pi\xi)\zeta(1+2\xi)\zeta(1-2\xi)} \times\{\cos(\tfrac{1}{2}\pi(u-v)) - \sin(\pi(\tfrac{1}{2}(z-w)+\xi))\}\Xi(\xi;u,v,w,z;g)d\xi; \tag{4.6.14}$$

and applying the functional equation (1.1.13) to $\zeta(1-2\xi)$ we get

$$\begin{aligned}&\mathscr{I}_{3,c}(u,v,w,z;g)\\ &= 2i(2\pi)^{w-z-2}\int_{(0)}(2\pi)^{2\xi}\{\cos(\tfrac{1}{2}\pi(u-v)) - \sin(\pi(\tfrac{1}{2}(z-w)+\xi))\}\\ &\times S(\xi;u,v,w,z)\Gamma(1-2\xi)\{\zeta(2\xi)\zeta(1+2\xi)\}^{-1}\Xi(\xi;u,v,w,z;g)d\xi.\end{aligned} \tag{4.6.15}$$

We then choose a Q which is to satisfy the condition

$$3B < Q \leq \tfrac{1}{4}A; \quad \zeta(s) \neq 0 \quad \text{for } \operatorname{Im} s = \pm Q.$$

We divide the range of integration in (4.6.15) into two parts according as $|\xi| > Q$ and $|\xi| \leq Q$, and denote the corresponding parts of $\mathscr{I}_{3,c}$ by $\mathscr{I}_{3,c}^{(1)}$ and $\mathscr{I}_{3,c}^{(2)}$, respectively. We observe that if $\operatorname{Re}\xi = 0$, $|\operatorname{Im}\xi| \geq Q$, then $S(\xi;u,v,w,z)$ is regular and $O(|\xi|^{cB})$ uniformly in $\mathscr{B}$ with an absolute constant c. Then, by virtue of Lemma 4.7, the integrand in the part $\mathscr{I}_{3,c}^{(1)}$ is regular and of fast decay with respect to ξ uniformly in $\mathscr{B}$. Hence $\mathscr{I}_{3,c}^{(1)}$ is regular in $\mathscr{B}$. As to $\mathscr{I}_{3,c}^{(2)}$ we move the path to L_Q which is the result of connecting the points $-Qi$, $[Q]+\frac{1}{4}-Qi$, $[Q]+\frac{1}{4}+Qi$, Qi with straight lines. The singularities of the integrand which we encounter in this procedure are all poles, and located at

$$\tfrac{1}{2}(u+v+w+z-3), \quad \tfrac{1}{2}(u+z-v-w-1), \quad \tfrac{1}{2}(v+z-u-w-1); \tag{4.6.16}$$

$$\tfrac{1}{2}\rho \quad (|\operatorname{Im}\rho| < Q); \quad \tfrac{1}{2}n \quad (2 \leq n \leq [Q]); \tag{4.6.17}$$

where ρ is a complex zero of $\zeta(s)$; note our choice of Q. The first three come from $S(\xi; u, v, w, z)$, and the others from $\Gamma(1-2\xi)\zeta(2\xi)^{-1}$, since we have here $(u, v, w, z) \in \mathscr{E}$ and so the Ξ-factor is regular for $\operatorname{Re}\xi \geq 0$. We may suppose, for an obvious reason, that the poles given in (4.6.16) are all simple, and do not coincide with any of those given in (4.6.17). Then we have

$$\mathscr{J}_{3,c}(u, v, w, z; g) = F_-(u, v, w, z; g) + U(u, v, w, z; g) + \mathscr{J}_{3,c}^{(Q)}(u, v, w, z; g). \tag{4.6.18}$$

Here F_- and U are the contributions of residues at the poles given in (4.6.16) and (4.6.17), respectively; and $\mathscr{J}_{3,c}^{(Q)}$ is the same as (4.6.15) but with the path $L_Q^\bullet$ which is the sum of the path L_Q and the half lines $(-i\infty, -Qi]$, $[Qi, i\infty)$. By virtue of Lemma 4.7, the terms F_- and U are meromorphic over $\mathscr{B}$, and $\mathscr{J}_{3,c}^{(Q)}$ is regular there.

Summing up the above discussion we have now proved

Lemma 4.9 *The function $\mathscr{J}_1(u, v, w, z; g)$ continues meromorphically to the domain $\mathscr{B}$. Thus the decomposition* (4.3.6) *holds throughout $\mathscr{B}$.*

4.7 Explicit formula

Having proved Lemma 4.9, it remains for us only to specialize (4.3.6) by setting $(u, v, w, z) = P_{\frac{1}{2}}$. This amounts to studying the local behavior, around $P_{\frac{1}{2}}$, of the various components of $\mathscr{J}_1$ which have been introduced in the above discussion. As a consequence we shall obtain the explicit formula for $\mathscr{Z}_2(g)$:

Theorem 4.2 *If g satisfies the basic assumption, then we have*

$$\mathscr{Z}_2(g) = \left\{\mathscr{Z}_{2,r} + \mathscr{Z}_{2,d} + \mathscr{Z}_{2,c} + \mathscr{Z}_{2,h}\right\}(g), \tag{4.7.1}$$

where

$$\mathscr{Z}_{2,r}(g) = \int_{-\infty}^{\infty} \sum_{\substack{a,b,k,l\geq 0\\ ak+bl\leq 4}} \varpi(a,k;b,l) \operatorname{Re}\left[\left(\frac{\Gamma^{(a)}}{\Gamma}\right)^k \left(\frac{\Gamma^{(b)}}{\Gamma}\right)^l (\tfrac{1}{2}+it)\right] g(t)dt$$
$$-2\pi \operatorname{Re}\left\{(c_E - \log(2\pi))g(\tfrac{1}{2}i) + \tfrac{1}{2} i g'(\tfrac{1}{2}i)\right\}$$

with effectively computable real absolute constants $\varpi(a,k;b,l)$; *and*

$$\mathscr{Z}_{2,d}(g) = \sum_{j=1}^{\infty} \alpha_j H_j(\tfrac{1}{2})^3 \Lambda(\kappa_j; g),$$

$$\mathscr{Z}_{2,c}(g) = \pi^{-1} \int_{-\infty}^{\infty} \frac{|\zeta(\frac{1}{2}+ir)|^6}{|\zeta(1+2ir)|^2} \Lambda(r;g) dr,$$

$$\mathscr{Z}_{2,h}(g) = \sum_{k=1}^{\infty} \sum_{j=1}^{\vartheta(2k)} \alpha_{j,2k} H_{j,2k}(\tfrac{1}{2})^3 \Lambda((\tfrac{1}{2}-2k)i; g).$$

Here c_E *is the Euler constant, and*

$$\begin{aligned}\Lambda(r;g) = \int_0^{\infty} &(y(1+y))^{-\frac{1}{2}} g_c(\log(1+1/y)) \\ &\times \mathrm{Re}\left[y^{-\frac{1}{2}-ir} \left(1 + \frac{i}{\sinh(\pi r)}\right) \frac{\Gamma(\frac{1}{2}+ir)^2}{\Gamma(1+2ir)} \right. \\ &\qquad \left. \times F(\tfrac{1}{2}+ir, \tfrac{1}{2}+ir; 1+2ir; -1/y) \right] dy \qquad (4.7.2)\end{aligned}$$

with the hypergeometric function F and

$$g_c(x) = \int_{-\infty}^{\infty} g(t) \cos(xt) dt.$$

Proof Combining (4.3.15), (4.5.3), (4.5.7), (4.6.18), and Lemma 4.9, we have the decomposition, over $\mathscr{B}$,

$$\mathscr{I}_1(u,v,w,z;g) = \{\mathscr{I}_2 + \mathscr{I}_{3,c} + \mathscr{I}_{3,d}^{+} + \mathscr{I}_{3,d}^{-} + \mathscr{I}_{3,h}^{+}\}(u,v,w,z;g) \quad (4.7.3)$$

with an obvious abuse of notation. In this identity $\mathscr{I}_{3,d}^{\pm}$ are regular at $P_{\frac{1}{2}}$. To see this we observe that when (u,v,w,z) is near $P_{\frac{1}{2}}$ the point (ir,u,v,w,z), with an arbitrary real r, is not in the set $\mathscr{N}$ defined by (4.6.2); thus by (4.4.19) and (4.4.20) the functions $\Phi_{\pm}(ir;u,v,w,z)$ are also regular at $P_{\frac{1}{2}}$ for any real r. Hence $\mathscr{I}_{3,d}^{\pm}$ are regular at $P_{\frac{1}{2}}$. Similarly we can see that $\mathscr{I}_{3,h}^{+}$ is regular at $P_{\frac{1}{2}}$. That is, we may set $(u,v,w,z) = P_{\frac{1}{2}}$ in the series expansions (4.5.5), (4.5.6), and (4.5.9) without any modification, and find that

$$\begin{aligned}&\{\mathscr{I}_{3,d}^{+} + \mathscr{I}_{3,d}^{-} + \mathscr{I}_{3,h}^{+}\}(P_{\frac{1}{2}}; g) \\ &= \sum_{j=1}^{\infty} \alpha_j H_j(\tfrac{1}{2})^3 \{\Phi_+ + \Phi_-\}(i\kappa_j; P_{\frac{1}{2}}; g) \\ &+ \sum_{k=1}^{\infty} \sum_{j=1}^{\vartheta(2k)} \alpha_{j,2k} H_{j,2k}(\tfrac{1}{2})^3 \Xi(2k-\tfrac{1}{2}; P_{\frac{1}{2}}; g) \qquad (4.7.4)\end{aligned}$$

with the conventions (4.4.1)–(4.4.3). It should be remarked that we have dropped ε_j's, because of (3.3.5); and the factor $(-1)^k$ has been eliminated, for $H_{j,k}(\frac{1}{2}) = 0$ if k is odd, which follows from (3.2.21). We note also that by (4.4.19) and (4.4.20) we have, for real r,

$$\{\Phi_+ + \Phi_-\}(ir; P_{\frac{1}{2}}; g) = \frac{1}{4}\Big(1 + \frac{i}{\sinh(\pi r)}\Big)\Xi(ir; P_{\frac{1}{2}}; g) + \frac{1}{4}\Big(1 - \frac{i}{\sinh(\pi r)}\Big)\Xi(-ir; P_{\frac{1}{2}}; g). \qquad (4.7.5)$$

On the other hand, Lemma 4.8 gives

$$\Xi(ir; P_{\frac{1}{2}}; g) = \frac{\Gamma(\frac{1}{2} + ir)^2}{\Gamma(1 + 2ir)} \int_0^\infty y^{-\frac{1}{2}+ir}(1 + y)^{-\frac{1}{2}} g^*(\log(1 + y)) \times F(\tfrac{1}{2} + ir, \tfrac{1}{2} + ir; 1 + 2ir; -y)dy. \qquad (4.7.6)$$

Hence

$$\{\Phi_+ + \Phi_-\}(ir; P_{\frac{1}{2}}; g) = \frac{1}{2}\int_0^\infty (y(1 + y))^{-\frac{1}{2}} g^*(\log(1 + y)) \times \operatorname{Re}\Big[y^{ir}\Big(1 + \frac{i}{\sinh(\pi r)}\Big)\frac{\Gamma(\frac{1}{2} + ir)^2}{\Gamma(1 + 2ir)} F(\tfrac{1}{2} + ir, \tfrac{1}{2} + ir; 1 + 2ir; -y)\Big] dy. \qquad (4.7.7)$$

Further, we observe that (4.7.6) holds with ir replaced by $k - \frac{1}{2}$; and thus the last identity gives, for any integer $k \geq 1$,

$$\{\Phi_+ + \Phi_-\}(2k - \tfrac{1}{2}; P_{\frac{1}{2}}; g) = \Xi(2k - \tfrac{1}{2}; P_{\frac{1}{2}}; g). \qquad (4.7.8)$$

Next, we consider $\mathscr{I}_{3,c}$ in the vicinity of $P_{\frac{1}{2}}$. We return to (4.6.18), and move the contour in $\mathscr{I}_{3,c}^{(Q)}$ back to the imaginary axis, while keeping (u, v, w, z) close to $P_{\frac{1}{2}}$. The poles which we encounter in this process are those given in (4.6.17) and $\frac{1}{2}(3 - u - v - w - z)$, which is close to $\frac{1}{2}$. For other poles of $S(\xi; u, v, w, z)$ are either close to $-\frac{1}{2}$ or canceled out by the zeros of the factor $\cos(\frac{1}{2}\pi(u - v)) - \sin(\pi(\frac{1}{2}(z - w) + \xi))$, and moreover Lemma 4.7 implies that $\Xi(\xi; u, v, w, z; g)$ is regular for $\operatorname{Re}\xi \geq -\frac{1}{4}$. We denote by $F_+(u, v, w, z; g)$ the contribution of the pole $\frac{1}{2}(3-u-v-w-z)$.

Then we have

$$\mathscr{I}^{(Q)}_{3,c}(u,v,w,z;g) = F_+(u,v,w,z;g) - U(u,v,w,z;g) + \mathscr{I}^*_{3,c}(u,v,w,z;g),$$

where $\mathscr{I}^*_{3,c}$ has the same expression as the right side of (4.6.15) but with different (u,v,w,z). Hence, by (4.6.18),

$$\mathscr{I}_{3,c}(u,v,w,z;g) = \{F_+ + F_-\}(u,v,w,z;g) + \mathscr{I}^*_{3,c}(u,v,w,z;g) \qquad (4.7.9)$$

when (u,v,w,z) is close to $P_{\frac{1}{2}}$. Here we should note that $\mathscr{I}^*_{3,c}$ is regular at $P_{\frac{1}{2}}$, and

$$\mathscr{I}^*_{3,c}(P_{\frac{1}{2}};g) = \frac{1}{\pi}\int_{-\infty}^{\infty} \frac{|\zeta(\frac{1}{2}+it)|^6}{|\zeta(1+2it)|^2}\{\Phi_+ + \Phi_-\}(it;P_{\frac{1}{2}};g)dt. \qquad (4.7.10)$$

This ends the local study of the decomposition (4.7.3) in the vicinity of $P_{\frac{1}{2}}$.

Now, invoking Lemma 4.9, we combine (4.3.6), (4.3.7), (4.7.3), and (4.7.9): If (u,v,w,z) is close to $P_{\frac{1}{2}}$, then we have

$$\begin{aligned}\mathscr{I}(u,v,w,z;g) &= \mathscr{M}(u,v,w,z;g) + \mathscr{I}^*_{3,c}(u,v,w,z;g)\\ &\quad + \overline{\mathscr{I}^*_{3,c}(\overline{w},\overline{z},\overline{u},\overline{v};g)} + \{\mathscr{I}^-_{3,d} + \mathscr{I}^+_{3,d} + \mathscr{I}^+_{3,h}\}(u,v,w,z;g)\\ &\quad + \overline{\{\mathscr{I}^-_{3,d} + \mathscr{I}^+_{3,d} + \mathscr{I}^+_{3,h}\}(\overline{w},\overline{z},\overline{u},\overline{v};g)}, \qquad (4.7.11)\end{aligned}$$

where

$$\begin{aligned}\mathscr{M}(u,v,w,z;g) &= \mathscr{I}_0(u,v,w,z;g) + \mathscr{I}_2(u,v,w,z;g)\\ &\quad + \overline{\mathscr{I}_2(\overline{w},\overline{z},\overline{u},\overline{v};g)} + \{F_+ + F_-\}(u,v,w,z;g)\\ &\quad + \overline{\{F_+ + F_-\}(\overline{w},\overline{z},\overline{u},\overline{v};g)}. \qquad (4.7.12)\end{aligned}$$

It should be stressed that all terms in (4.7.11) are regular at $P_{\frac{1}{2}}$. That the function $\mathscr{M}$ is regular at $P_{\frac{1}{2}}$ is due to the fact that all terms in (4.7.11) except for $\mathscr{M}$ have already been proved to be regular at $P_{\frac{1}{2}}$.

In view of (4.7.4) and (4.7.10) it remains for us to express $\mathscr{M}(P_{\frac{1}{2}};g)$ in terms of g, which is a somewhat complicated task. For this purpose we compute $F_\pm$ explicitly. Then we have

$$\mathscr{M}(u,v,w,z;g) = \sum_{j=0}^{12} \mathscr{M}_j(u,v,w,z;g)$$

with

$$\mathscr{M}_{6+j}(u,v,w,z;g) = \overline{\mathscr{M}_j(\overline{w},\overline{z},\overline{u},\overline{v};g)} \quad (1 \le j \le 6).$$

Here

$$\mathcal{M}_0(u,v,w,z) = g^*(0)\zeta(u+w)\zeta(u+z)\zeta(v+w)\zeta(v+z)\{\zeta(u+v+w+z)\}^{-1},$$

$$\mathcal{M}_1(u,v,w,z) = \tilde{g}(v+w-1,w)\zeta(u+z)\zeta(v+w-1)\zeta(z-w+1) \times \zeta(u-v+1)\{\zeta(u+z-v-w+2)\}^{-1},$$

$$\mathcal{M}_2(u,v,w,z) = \tilde{g}(u+w-1,w)\zeta(v+z)\zeta(u+w-1)\zeta(z-w+1) \times \zeta(v-u+1)\{\zeta(v+z-u-w+2)\}^{-1},$$

$$\begin{aligned}\mathcal{M}_3(u,v,w,z;g) = (2\pi)^{w-z}&\{\cos(\tfrac{1}{2}\pi(u-v))+\cos(\pi(z-w+\tfrac{1}{2}(u-v)))\}\\ &\times \zeta(u+z-1)\zeta(v+w)\zeta(z-w)\zeta(v-u+1)\\ &\times \{\cos(\tfrac{1}{2}\pi(u+z-v-w))\zeta(2-u-z+v+w)\}^{-1}\\ &\times \Xi(\tfrac{1}{2}(u+z-v-w-1);u,v,w,z;g),\end{aligned}$$

$$\begin{aligned}\mathcal{M}_4(u,v,w,z;g) = (2\pi)^{w-z}&\{\cos(\tfrac{1}{2}\pi(u-v))+\cos(\pi(z-w+\tfrac{1}{2}(v-u)))\}\\ &\times \zeta(v+z-1)\zeta(u+w)\zeta(z-w)\zeta(u-v+1)\\ &\times \{\cos(\tfrac{1}{2}\pi(v+z-u-w))\zeta(2-v-z+u+w)\}^{-1}\\ &\times \Xi(\tfrac{1}{2}(v+z-u-w-1);u,v,w,z;g),\end{aligned}$$

$$\begin{aligned}\mathcal{M}_5(u,v,w,z;g) = -(2\pi)^{w-z}&\{\cos(\tfrac{1}{2}\pi(u-v))-\cos(\pi(z+\tfrac{1}{2}(u+v)))\}\\ &\times \zeta(u+z-1)\zeta(2-v-w)\zeta(v+z-1)\zeta(2-u-w)\\ &\times \{\cos(\tfrac{1}{2}\pi(u+v+w+z))\zeta(4-u-v-w-z)\}^{-1}\\ &\times \Xi(\tfrac{1}{2}(u+v+w+z-3);u,v,w,z;g),\end{aligned}$$

$$\begin{aligned}\mathcal{M}_6(u,v,w,z;g) = (2\pi)^{w-z}&\{\cos(\tfrac{1}{2}\pi(u-v))+\cos(\pi(w+\tfrac{1}{2}(u+v)))\}\\ &\times \zeta(u+z-1)\zeta(2-v-w)\zeta(v+z-1)\zeta(2-u-w)\\ &\times \{\cos(\tfrac{1}{2}\pi(u+v+w+z))\zeta(4-u-v-w-z)\}^{-1}\\ &\times \Xi(-\tfrac{1}{2}(u+v+w+z-3);u,v,w,z;g).\end{aligned}$$

Among these, $\mathcal{M}_0$ is equal to $\mathcal{J}_0$; $\mathcal{M}_1$ and $\mathcal{M}_2$ come from $\mathcal{J}_2$; and $\mathcal{M}_j$ $(3 \le j \le 6)$ are the contributions of residues of the integral in (4.6.14) at the poles $\xi = \frac{1}{2}(u+z-v-w-1)$, $\frac{1}{2}(v+z-u-w-1)$, $\frac{1}{2}(u+v+w+z-3)$, $\frac{1}{2}(3-u-v-w-z)$, respectively. They can be singular at $P_{\frac{1}{2}}$ individually, but the singular parts should cancel each other out if they are brought into (4.7.12), for $\mathcal{M}$ is regular at $P_{\frac{1}{2}}$. More precisely, put $(u,v,w,z) = P_{\frac{1}{2}} + (a_1,a_2,a_3,a_4)\delta$ with a small complex δ, and expand each term into a Laurent series in δ; then the sum of the constant terms is equal to $\mathcal{M}(P_{\frac{1}{2}})$, regardless of the choice of the vector (a_1, a_2, a_3, a_4). We choose it in such a way that it is real and no singularities of any of the $\mathscr{M}_j$ $(0 \le j \le 13)$ are encountered when $|\delta|$ tends to 0. This is possible, for the exceptional a_1, a_2, a_3, a_4 satisfy a finite number of linear relations. Thus we shall assume hereafter that $\delta \neq 0$ is small and the vector (a_1, a_2, a_3, a_4) is chosen accordingly; and we denote $(a_1, a_2, a_3, a_4)\delta$ either by (δ) or by $(\delta_1, \delta_2, \delta_3, \delta_4)$. Also we denote the constant term of $\mathscr{M}_j$ by $\mathscr{M}_j^*$.

Since $\mathscr{M}_0$ is trivial we begin with $\mathscr{M}_1$. To it we apply (4.1.11), and have

$$\begin{aligned}\mathscr{M}_1(P_{\frac{1}{2}} + (\delta)) &= \Gamma(\delta_2 + \delta_3)\zeta(\delta_1 + \delta_4 + 1)\zeta(\delta_4 - \delta_3 + 1)\zeta(\delta_1 - \delta_2 + 1) \\ &\quad \times \zeta(\delta_2 + \delta_3)\{\zeta(2 + \delta_1 - \delta_2 - \delta_3 + \delta_4)\}^{-1} \\ &\quad \times \int_{-\infty}^{\infty} \frac{\Gamma(\frac{1}{2} - \delta_2 - it)}{\Gamma(\frac{1}{2} + \delta_3 - it)} g(t) dt.\end{aligned}$$

This implies that the singularity of $\mathscr{M}_1$ at $P_{\frac{1}{2}}$ is of order 4. Hence the constant term of $\mathscr{M}_1(P_{\frac{1}{2}} + (\delta))$ is a linear combination of the first five coefficients of the power series in δ for the last integral. Thus

$$\mathscr{M}_1^* = \int_{-\infty}^{\infty} \sum_{\substack{a,b,k,l \ge 0 \\ ak+bl \le 4}} d(a,k;b,l) \left(\frac{\Gamma^{(a)}}{\Gamma}\right)^k \left(\frac{\Gamma^{(b)}}{\Gamma}\right)^l (\tfrac{1}{2} - it) g(t) dt \tag{4.7.13}$$

with an obvious abuse of notation; the constants $d(a, b; k, l)$ are real and may depend on (a_1, a_2, a_3, a_4). Clearly $\mathscr{M}_2$ can be treated in just the same way, and $\mathscr{M}_2^*$ has the same form as (4.7.13).

The terms $\mathscr{M}_j$ $(3 \le j \le 6)$ are not so simple; and our computation of them depends on the formula (2.6.13). By the definition (4.4.18) we have, for the Ξ-factor in $\mathscr{M}_3$,

$$\begin{aligned}&\Xi(\tfrac{1}{2}(u + z - v - w - 1); u, v, w, z; g) \\ &\quad = \frac{1}{2\pi i} \int_{-i\infty}^{i\infty} \Gamma(u + z - 1 - s)\Gamma(s + 1 - u - w)\tilde{g}(s, w) ds, \end{aligned} \tag{4.7.14}$$

where the path separates the poles of $\Gamma(u + z - 1 - s)$ and those of the other two factors to the right and the left, respectively; that we can draw such a path is assured by our choice of (a_1, a_2, a_3, a_4). Inserting (4.1.11) in this we get an absolutely convergent double integral; and thus we have

$$\begin{aligned}
&\Xi(\tfrac{1}{2}(u+z-v-w-1);u,v,w,z;g)\\
&\quad=\int_{-\infty}^{\infty}\frac{g(t)}{\Gamma(w-it)}\\
&\quad\times\frac{1}{2\pi i}\int_{-i\infty}^{i\infty}\Gamma(s)\Gamma(s+1-u-w)\Gamma(u+z-1-s)\Gamma(w-it-s)dsdt.
\end{aligned}$$

The path of the inner integral is the same as in (4.7.14); and obviously we may suppose that it separates the poles of the first two Γ-factors from those of the other two. Hence we have, by (2.6.13),

$$\begin{aligned}
&\Xi(\tfrac{1}{2}(u+z-v-w-1);u,v,w,z;g)\\
&\qquad=\Gamma(u+z-1)\Gamma(z-w)\int_{-\infty}^{\infty}\frac{\Gamma(1-u-it)}{\Gamma(z-it)}g(t)dt.
\end{aligned}$$

This implies that $\mathcal{M}_3$ has a singularity of order 4 at $P_{\frac{1}{2}}$; thus $\mathcal{M}_3^*$ admits an expression of the same form as (4.7.13). Obviously the same argument applies to $\mathcal{M}_4$.

We next consider $\mathcal{M}_5$. Its Ξ-factor is, as in the case of $\mathcal{M}_2$,

$$\begin{aligned}
&\Xi(\tfrac{1}{2}(u+v+w+z-3);u,v,w,z;g)\\
&\quad=\int_{-\infty}^{\infty}\frac{g(t)}{\Gamma(w-it)}\frac{1}{2\pi i}\int_{-i\infty}^{i\infty}\Gamma(s+1-u-w)\Gamma(s+1-v-w)\\
&\quad\times\Gamma(u+v+w+z-2-s)\Gamma(w-it-s)dsdt,
\end{aligned}$$

where the path is in accord with (2.6.13). Thus we have

$$\begin{aligned}
&\Xi(\tfrac{1}{2}(u+v+w+z-3);u,v,w,z;g)\\
&\quad=\Gamma(u+z-1)\Gamma(v+z-1)\int_{-\infty}^{\infty}\frac{\Gamma(1-u-it)\Gamma(1-v-it)}{\Gamma(w-it)\Gamma(z-it)}g(t)dt.
\end{aligned}$$

This implies again that $\mathcal{M}_5^*$ has an expression of the same form as (4.7.13).

We then move to $\mathcal{M}_6$. This is regular at $P_{\frac{1}{2}}$. For we have

$$\begin{aligned}
&\cos(\tfrac{1}{2}\pi(u-v))+\cos(\pi(w+\tfrac{1}{2}(u+v)))\\
&\qquad=2\sin(\tfrac{1}{2}\pi(u+w-1))\sin(\tfrac{1}{2}\pi(v+w-1)),
\end{aligned}$$

which cancels out the singularities of $\zeta(2-u-w)\zeta(2-v-w)$; and moreover the Ξ-factor there is regular at $P_{\frac{1}{2}}$ because of Lemma 4.7. Thus we have

$$\mathcal{M}_6^*=-\frac{3}{4}\Xi(\tfrac{1}{2};P_{\frac{1}{2}};g).$$

To compute this Ξ-factor we may use Lemma 4.8 and the fact that $F(1,1;2;-y) = y^{-1}\log(1+y)$ for $y \geq 0$. But we shall show an alternative argument which depends on (2.6.13). By (4.1.11) and (4.4.18) we have

$$\Xi(\tfrac{1}{2}; P_{\frac{1}{2}}; g) = -\int_{-\infty}^{\infty} \frac{g(t)}{\Gamma(\frac{1}{2}-it)} C(t)dt,$$

where

$$C(t) = \frac{1}{2\pi i}\int_{(\frac{1}{4})} \Gamma(s)^2\Gamma(-s)\Gamma(\tfrac{1}{2}-it-s)ds.$$

We have

$$\begin{aligned} C(t) &= \lim_{\alpha\to 0+} \frac{1}{2\pi i}\int_{(\frac{1}{4})} \Gamma(s)^2\Gamma(\alpha-s)\Gamma(\tfrac{1}{2}-it-s)ds \\ &= \lim_{\alpha\to 0+}\left\{\frac{1}{2\pi i}\int_{(\frac{1}{2}\alpha)} \Gamma(s)^2\Gamma(\alpha-s)\Gamma(\tfrac{1}{2}-it-s)ds - \Gamma(\alpha)^2\Gamma(\tfrac{1}{2}-it-\alpha)\right\}. \end{aligned}$$

To the last integral we can apply (2.6.13); so we get

$$\begin{aligned} C(t) &= \Gamma(\tfrac{1}{2}-it)\lim_{\alpha\to 0+}\Gamma(\alpha)^2\left\{\frac{\Gamma(\frac{1}{2}-it)}{\Gamma(\frac{1}{2}-it+\alpha)} - \frac{\Gamma(\frac{1}{2}-it-\alpha)}{\Gamma(\frac{1}{2}-it)}\right\} \\ &= -\Gamma(\tfrac{1}{2}-it)\left(\frac{\Gamma'}{\Gamma}\right)'(\tfrac{1}{2}-it). \end{aligned}$$

This implies that $\mathcal{M}_6^*$, too, is of the same form as (4.7.13); and we end our discussion on $\mathcal{M}(P_{\frac{1}{2}})$.

Finally, combining these and Lemma 4.2, (4.7.4), (4.7.7), (4.7.8), (4.7.10), and (4.7.11), we end the proof of Theorem 4.2.

4.8 Notes for Chapter 4

Readers who want to have an overview of the general theory as well as the history of the mean-value problem of the zeta-function are referred to Ivić [20]. See also the survey article [52].

The explicit formula (4.7.1) appeared first in Motohashi [47] but in a less sophisticated form. The older version is the specialization of the present version with g being replaced by the Gaussian weight (see the next chapter). Moreover, it contained a remainder term, which has been eliminated in the above. This is due to a more effective treatment of the main term $\mathcal{M}(P_{\frac{1}{2}})$. Otherwise not many changes have been added. The formula (4.7.1) should be compared with (4.1.16), which is a variant of Atkinson's important work [2] (see Jutila [30] and Motohashi [45] for alternative approaches to Atkinson's result). The similarities between these two formulas are rather striking. The terms $\mathscr{Z}_{2,r}$ and $\mathscr{Z}_{2,d}$ are clearly of the same type as the corresponding parts of (4.1.16). The arithmetic part in the spectral series in (4.1.16), i.e., the divisor function, is replaced here by the special value of the Hecke L-function, though cubed, which is naturally an arithmetic quantity. Also the cosine function in (4.1.16) is replaced essentially by the hypergeometric function. They are related with resolvent kernels in the relevant theories of harmonic analysis: Behind the former is Poisson's sum formula, and behind the latter Kuznetsov's trace formula, which is a non-Euclidean counterpart of the former. Indeed the function $\Lambda(r;g)$ defined by (4.7.2) is closely related with the free-space resolvent kernel $r_\alpha(z,w)$ defined by (1.1.49).

The contribution of the continuous spectrum contains the sixth power of the zeta-function, which is somewhat annoying and ironical, since we have encountered something more difficult while investigating the fourth power moment. This is, however, superficial; and in fact $\mathscr{Z}_{2,c}(g)$ is in general quite small because of the classical bounds and the rapid decay of $\Lambda(r;g)$. The same can be said about $\mathscr{Z}_{2,h}(g)$. Thus the analytical peculiarities of $\mathscr{Z}_2(g)$ are essentially controlled by the discrete spectrum of the non-Euclidean Laplacian over the full modular group. These facts will be made more perceptible in the next chapter.

Another feature of (4.7.1) that we cannot fail to observe is its similarity to the explicit formula in the theory of prime numbers. The rôle played by the complex zeros of $\zeta(s)$ in the latter formula is analogous here to that played by the discrete eigenvalues. Moreover, from this viewpoint it appears that the trivial zeros of $\zeta(s)$ correspond to holomorphic cusp-forms. Their contributions to the respective formulas are comparable; in fact both are negligible in general. Naturally, behind this fascinating story is the zeta-function of Selberg. But it looks hard, at least presently, to establish any direct linkage between our explicit formula and Selberg's zeta-function. Somewhat informally expressing it, these matters appear to suggest to us that we should reconstruct the whole theory starting from the automorphic resolvent kernel $R_\alpha(z,w)$ defined by (1.1.51). The geometric computation of $R_\alpha(z,w)$ leads us either to Selberg's or to Kuznetsov's trace formula. Using the latter we have proved (4.7.1). Thus it is reasonable for us to suppose that there would be a way to integrate $R_\alpha(z,w)$ to produce an expression, the Fourier expansion of which involves, on the geometric side, the fourth power moment of the zeta-function and, on the spectral side, Hecke L-functions. Then the relation between (4.7.1) and Selberg's zeta-function would become clear, since the whole mechanism is centered at the resolvent kernel. At any event the formula (4.7.1) illustrates a peculiar position of the Riemann zeta-function in the entire world of automorphic L-functions. The zeta-function is served by all automorphic L-functions!

It appears that the phenomenon observed in Section 4.5 and consequently the formula (4.7.1) have a certain generality. For they extend to a situation where the underlying group is not the full modular group: We consider

$$\mathscr{Z}(g,\chi) = \int_{-\infty}^{\infty} |\zeta(\tfrac{1}{2}+it)L(\tfrac{1}{2}+it,\chi)|^2 g(t)dt,$$

where χ is supposed to be a primitive Dirichlet character mod q with $q > 1$. This includes the mean square of Dedekind zeta-functions of quadratic number fields (see Motohashi [56]). A modification of the argument in Section 4.3 leads us to sums of Kloosterman sums whose moduli are divisible by q, i.e., $S(m,n;ql)$. On the other hand the Fourier coefficients of the Poincaré series at the cusp $i\infty$ of the congruence subgroup $\Gamma_0(q)$ (cf. (3.1.11)) involves these trigonometrical sums in just the same way as (1.1.6) does. Thus trace formulas for the non-Euclidean Laplacian over $\Gamma_0(q)$ are to be developed. As was mentioned already, this does not cause any more difficulties than to follow closely the analysis of Chapters 1–3, or rather just to refer to Deshouillers and Iwaniec [8]. Here there is, however, a problem related to the Hecke operators, since the theory given in Section 3.1 does not extend to $\Gamma_0(q)$ straightforwardly. Nevertheless, as far as the application to $\mathscr{Z}(g,\chi)$ is concerned, there is not much to change in the argument of Section 4.1. For the presence of the character χ allows us to restrict the action of Hecke operators $T(n)$, $T_k(n)$ to those n with $(n,q)=1$ for which the definitions (3.1.3) and (3.1.19) are still valid. Then, using the reasoning of this chapter, we eventually end up with a spectral decomposition of $\mathscr{Z}(g,\chi)$. It is analogous to (4.7.1) and contains the factors $|L_j(\frac{1}{2})|^2 H_j(\frac{1}{2},\chi)$ in place of $\alpha_j H_j(\frac{1}{2})^3$. Here the Hecke L-function $L_j(s)$ is naturally the one attached to a cusp-form over $\Gamma_0(q)$; and $H_j(s,\chi)$ is the χ-twist of $L_j(s)$ with an obvious normalization. It should be worth remarking that if χ is real and $\chi(-1)=-1$ (i.e., the case of imaginary quadratic number fields) the explicit formula for $\mathscr{Z}(g,\chi)$ does not contain the contribution coming from holomorphic cusp-forms. This suggests that there exists an interaction between the zeta- and Dirichlet L-functions in a somewhat mysterious way.

Also, it is possible to consider the extension to the mean value problem of Iwaniec [23]:

$$\int_{-T}^{T} |\zeta(\tfrac{1}{2}+it)|^4 \Big|\sum_{N\le n\le 2N} a_n n^{it}\Big|^2 dt,$$

where $\{a_n\}$ is an arbitrary complex vector (see Deshouillers and Iwaniec [9]). This is essentially the same as to decompose spectrally the expression

$$\mathscr{Z}_2(g, b/a) = \int_{-\infty}^{\infty} |\zeta(\tfrac{1}{2}+it)|^4 (b/a)^{it} g(t) dt,$$

where $(a,b)=1$ (see Motohashi [55]). A straightforward use of (4.2.4) reduces this to the study of the sums of the form

$$\sum_{\substack{l\ge 1\\ (l,c)=1\\ l\equiv 0 \bmod d}} \frac{1}{l\sqrt{c}} S(m, \bar{c}n; l)\varphi\Big(\frac{4\pi\sqrt{mn}}{l\sqrt{c}}\Big),$$

where $c|a$, $d|b$. The point is that these Kloosterman sums can be produced via the inner-product of two Poincaré series over $\Gamma_0(cd)$. Thus $\mathscr{Z}_2(g, b/a)$ is related to congruence subgroups $\Gamma_0(q)$ such that $q|ab$; and the situation becomes similar to that of $\mathscr{Z}(g,\chi)$. There is, however, a notable difference, too. This concerns the way of achieving the analytic continuation of the intermediate spectral decomposition. The task itself is analogous to what we encountered above. But this time, unlike in the cases of $\mathscr{Z}_2(g)$ and $\mathscr{Z}(g,\chi)$, we are unable to enjoy the power of Hecke operators fully.

It is expedient to see the situation more precisely, since it will reveal that the formulas (4.5.5), (4.5.6), and (4.5.9) are in fact fortuitous outcomes, and that an essential ramification of the line of reasoning is inevitable: The trouble is caused by the function

$$\mathfrak{D}(s,\alpha) = \sum_{n=1}^{\infty} \rho(n)\sigma_\alpha(n) n^{-s},$$

where $\rho(n)$'s are the Fourier coefficients around the point at $i\infty$ of a cusp-form ψ over $\Gamma_0(q)$; and s, α are independent complex variables. We can assume that ψ is an eigenfunction of all Hecke operators $T(n)$ with $(n,q)=1$, where $T(n)$ is defined by (3.1.3). What we need is a meromorphic continuation of $\mathfrak{D}(s,\alpha)$ to a sufficiently wide domain while specifying the location of singularities. If ψ is a new-form in the sense of Atkin and Lehner [1] then we get

$$\mathfrak{D}(s,\alpha) = \rho(1)H(s)H(s-\alpha)/\zeta_q(2s-\alpha)$$

with $\zeta_q(s) = \zeta(s)\prod_{p|q}(1-p^{-s})$, which is precisely an analogue of (3.2.7). Thus $\zeta_q(2s-\alpha)\mathfrak{D}(s,\alpha)$ is entire over $\mathbb{C}^2$, and this case yields no trouble. If ψ is an old-form, we still have an identity similar to the above; and $\zeta_q(2s-\alpha)\mathfrak{D}(s,\alpha)$ is again entire over $\mathbb{C}^2$. But in general it can be said only that there is a new-form ψ^* over a $\Gamma_0(r)$, $r|q$, such that if $T(n)\psi = \kappa\psi$ then $T(n)\psi^* = \kappa\psi^*$ for all n, $(n,q)=1$. This is not sufficient to deduce the desired analytic continuation of $\mathfrak{D}(s,\alpha)$, without which it is hard to proceed further.

Therefore we have to move to an entirely different argument, which we call the convolution method. This is of course applicable to the case of the full modular group, too; and we are able to prove in an alternative way the contents of Lemma 4.9 without recourse to Lemma 4.5. We stress that Lemma 4.9 is the key to Theorem 4.2. The convolution method can establish analogues of Lemma 4.9 for the problems $\mathscr{Z}(g,\chi)$ and $\mathscr{Z}_2(g,b/a)$ as well. Hence, in particular, $\mathscr{Z}_2(g,b/a)$ admits an explicit formula. In the context of the full modular group the convolution method was, as a matter of fact, employed already in the proof of Lemma 3.5. Thus it would be sufficient to note the relation

$$\mathfrak{D}(s,\alpha) = \frac{4\pi^{s-\frac{1}{2}\alpha}}{\Gamma(s-\frac{1}{2}\alpha;\alpha,i\kappa)}\int_{\mathscr{F}}\psi(z)E^*(z,\tfrac{1}{2}(1-\alpha))E(z,s-\tfrac{1}{2}\alpha)d\mu(z),$$

where the Γ-factor is defined by (3.2.12) and $E^*(z,s) = \pi^{-s}\Gamma(s)\zeta(2s) \times E(z,s)$. Here we have assumed ψ is even and corresponds to the eigenvalue $\kappa^2+\frac{1}{4}$. The odd case is analogous to (3.2.19). This expression yields readily that $\zeta(2s-\alpha)\mathfrak{D}(s,\alpha)$ is entire, which is exactly what we needed in the proof of Lemma 4.9. Thus we could dispense with Hecke's theory. However, comparing the outcomes we prefer using Hecke's theory in the case of the full modular group for an obvious reason.

Another interesting problem is to find an analogue of Theorem 4.2 for individual Dirichlet L-functions:

$$\int_{-\infty}^{\infty}|L(\tfrac{1}{2}+it,\chi)|^4 g(t)dt$$

with an arbitrary primitive χ. It is possible to reduce this to sums of Kloosterman sums. The relevant sum is

$$\sum_{a=1}^{q}\overline{\chi}(a)\chi(a+b)\sum_{\substack{k=1\\(k,q)=1}}^{\infty}\frac{1}{k}S(q^*m,\pm q^*bn;k)S(k^*m,\pm k^*an;q)\varphi(\frac{4\pi\sqrt{bmn}}{qk}),$$

where q is the conductor of χ; $b,m,n\geq 1$, $q^*q\equiv 1 \bmod k$, $k^*k\equiv 1 \bmod q$; and φ is similar to $\varphi_{\pm}$ above. The remaining problem, which is left for readers, is to produce these Kloosterman sums by a suitable combination of Poincaré series over a certain congruence group. For a statistical aspect of this mean value problem see Jutila and Motohashi [33].

Further, it is desirable to have an analogue of Theorem 4.2 for automorphic L-functions. The case of holomorphic cusp-forms is easy, and a solution is given in Motohashi [50]. The real interest is, however, in the case of Maass cusp-forms. This is not only because of possible applications to the theory of mean values of the zeta-function but also because the required method obviously needs to be structural. Approaches to this basic problem are developed in the works of Jutila listed in the bibliography.

Finally, we remark that there exists a possibility of extending the results of this chapter to Dedekind zeta-functions of imaginary quadratic number fields. This is because we have necessary trace formulas for the Picard–Bianchi groups as we indicated in Section 2.8. The interest lies in the fact that the relevant mean value of Dedekind zeta-functions is related to the eighth power moment of the Riemann zeta-function.

Here is a technical comment: The assertions (4.4.9) and (4.4.11) could be avoided in developing the spectral expansion of $K_+(m,n;u,v,w,z;g)$. In fact we may repeat the argument around (2.4.13) with the particular choice of φ, as was done in [47, p. 198]. This leads us to (4.4.12)–(4.4.13) relatively quickly. But the present argument is more in accord with the main context of this monograph.

For the sake of completeness we shall show the proofs of (4.4.9) and (4.4.11); for the published accounts (cf. [38], [74]) rely on the theory of Hankel functions, and are involved. We prove first, a little more generally than (4.4.9), the estimate

$$J_\nu(x) \ll \begin{cases} x^{\mathrm{Re}\,\nu} & \text{if } 0 < x \le 1, \\ x^{-\frac{1}{2}} & \text{if } x \ge 1, \end{cases}$$

provided $\mathrm{Re}\,\nu > -\frac{1}{2}$, where the implied constant depends on ν. Because of (2.2.8) it is enough to consider the estimation of

$$\int_{-1}^{1} (1-y^2)^{\nu-\frac{1}{2}} e^{ixy}\, dy.$$

Or, replacing y by $1-2u$, we see that we need to bound

$$\int_0^1 (u(1-u))^{\nu-\frac{1}{2}} e^{2ixu}\, du.$$

We may obviously restrict ourselves to the case where x is large. We then turn the path through $\frac{1}{2}\pi$ and use Cauchy's theorem. The circular part of the new path contributes

$$\ll \int_0^{\frac{1}{2}\pi} \theta^{\mathrm{Re}\,\nu-\frac{1}{2}} e^{-x\theta} d\theta \ll x^{-\mathrm{Re}\,\nu-\frac{1}{2}}.$$

The contribution of the segment on the positive imaginary axis is easily seen to be less than the same bound, which ends the proof of the above claim. Thus we see also that the integral in (4.4.11) is absolutely and uniformly convergent in the indicated range of ν and μ. We then restrict ourselves to the range $0 > \mathrm{Re}\,(\nu - \mu) > -1$, $\mathrm{Re}\,\mu > -\frac{1}{2}$, where we shall verify (4.4.11), and finish the proof by analytic continuation. On this condition we have, by (2.2.8),

$$\begin{aligned}
&\int_0^\infty J_\nu(x)\left(\frac{x}{2}\right)^{-\mu} dx \\
&\quad = \frac{2^{\mu-\nu+1}}{\sqrt{\pi}\Gamma(\nu+\frac{1}{2})} \lim_{X\to\infty} \int_0^X x^{\nu-\mu} \int_0^1 (1-y^2)^{\nu-\frac{1}{2}} \cos(xy) dy dx \\
&\quad = \frac{2^{\mu-\nu+1}}{\sqrt{\pi}\Gamma(\nu+\frac{1}{2})} \lim_{X\to\infty} \int_0^1 (1-y^2)^{\nu-\frac{1}{2}} y^{\mu-\nu-1} \int_0^{Xy} x^{\nu-\mu} \cos x \, dx dy.
\end{aligned}$$

The part corresponding to $y < X^{-1}$ clearly contributes $O(X^{\mathrm{Re}\,(\nu-\mu)})$. On the other hand, if $y \ge X^{-1}$, then we have

$$\int_0^{Xy} \cdots dx = \Gamma(\nu+1-\mu)\sin(\tfrac{1}{2}(\mu-\nu)\pi) - \int_{Xy}^\infty \cdots dx.$$

The last integral is $O((Xy)^{\mathrm{Re}\,(\nu-\mu)})$ by integration by parts, which contributes $O(X^{\mathrm{Re}\,(\nu-\mu)} \log X)$. Hence we get

$$\int_0^\infty J_\nu(x)\left(\frac{x}{2}\right)^{-\mu} dx = \frac{2^{\mu-\nu}}{\sqrt{\pi}} \frac{\Gamma(\frac{1}{2}(\mu-\nu))}{\Gamma(\frac{1}{2}(\nu+1+\mu))} \Gamma(\nu+1-\mu)\sin(\tfrac{1}{2}(\mu-\nu)\pi).$$

Applying the duplication formula to the factor $\Gamma(\nu+1-\mu)$ we obtain the identity (4.4.11).

5
Asymptotics

The aim of the present chapter is to crop some quantitative information about the zeta-function from the explicit formula (4.7.1). More specifically, we shall study the case $k = 2$ of

$$I_k(T,G) = (\sqrt{\pi}G)^{-1}\int_{-\infty}^{\infty} |\zeta(\tfrac{1}{2} + i(T+t))|^{2k} \exp(-(t/G)^2)dt \quad (T,\ G > 0).$$

If G is small compared with T, then supposedly $I_k(T,G)$ reflects well the peculiarities of $\zeta(s)$ around the point $\frac{1}{2} + iT$. We shall try to find an asymptotic expression for $I_2(T,G)$ that is sufficiently explicit in terms of the variables T, G. Our results exhibit an intrinsic relation between the zeta-values and the discrete spectrum of Δ over the full modular group, more perceptively than in (4.7.1).

5.1 Local explicit formula

Now we obviously have

$$I_2(T,G) = \mathscr{Z}_2(g)$$

with

$$g(t) = (\sqrt{\pi}G)^{-1}\exp(-((T-t)/G)^2);$$

and this g satisfies the basic assumption for any fixed A. We have, for its Fourier cosine transform g_c,

$$g_c(x) = \exp(-\tfrac{1}{4}(Gx)^2)\cos(Tx).$$

Then Theorem 4.2 gives, for any $T, G > 0$,

$$I_2(T,G) = \{I_{2,r} + I_{2,d} + I_{2,c} + I_{2,h}\}(T,G), \tag{5.1.1}$$

where

$$I_{2,r}(T,G) = (\sqrt{\pi}G)^{-1}\int_{-\infty}^{\infty}\sum_{\substack{a,b,k,l\ge 0\\ ak+bl\le 4}}\varpi(a,k;b,l)$$
$$\times \operatorname{Re}\left[\left(\frac{\Gamma^{(a)}}{\Gamma}\right)^k\left(\frac{\Gamma^{(b)}}{\Gamma}\right)^l(\tfrac12+i(T+t))\right]e^{-(t/G)^2}dt$$
$$-2\sqrt{\pi}G^{-1}\operatorname{Re}\left[\{c_E-\log(2\pi)+(\tfrac12+iT)G^{-2}\}\exp(((\tfrac12+iT)/G)^2)\right]$$

with certain absolute constants $\varpi(a,k;b,l)$; and

$$I_{2,d}(T,G)=\sum_{j=1}^{\infty}\alpha_jH_j(\tfrac12)^3\Lambda(\kappa_j;T,G);$$

$$I_{2,c}(T,G)=\pi^{-1}\int_{-\infty}^{\infty}\frac{|\zeta(\frac12+ir)|^6}{|\zeta(1+2ir)|^2}\Lambda(r;T,G)dr;$$

$$I_{2,h}(T,G)=\sum_{k=1}^{\infty}\sum_{j=1}^{\vartheta(2k)}\alpha_{j,2k}H_{j,2k}(\tfrac12)^3\Lambda((\tfrac12-2k)i\,;T,G)$$

with

$$\Lambda(r;T,G)$$
$$=\int_0^{\infty}(y(1+y))^{-\frac12}\cos(T\log(1+1/y))\exp(-\tfrac14(G\log(1+1/y))^2)$$
$$\times\operatorname{Re}\left[y^{-\frac12-ir}\left(1+\frac{i}{\sinh(\pi r)}\right)\frac{\Gamma(\frac12+ir)^2}{\Gamma(1+2ir)}\right.$$
$$\left.\times F(\tfrac12+ir,\tfrac12+ir;1+2ir;-1/y)\right]dy. \tag{5.1.2}$$

Thus our problem is essentially equivalent to finding the asymptotic expansion for $\Lambda(r;T,G)$. For this purpose we shall use (5.1.2) not in the present form but in its original form coming from (4.7.5) and (4.7.8); that is, we have, instead of (5.1.2),

$$\Lambda(r;T,G)$$
$$=\frac12\operatorname{Re}\left[\left(1+\frac{i}{\sinh(\pi r)}\right)\Xi(ir;T,G)+\left(1-\frac{i}{\sinh(\pi r)}\right)\Xi(-ir;T,G)\right] \tag{5.1.3}$$

if r is real, and

$$\Lambda((\tfrac12-2k)i;T,G)=2\operatorname{Re}\left[\Xi(2k-\tfrac12;T,G)\right] \tag{5.1.4}$$

if k is a positive integer. Here we have, by the definition (4.4.18),

$$\Xi(\xi; T, G) = \frac{1}{2\pi i}\int_{(\frac{1}{4})} \frac{\Gamma(\xi + \frac{1}{2} - s)}{\Gamma(\xi + \frac{1}{2} + s)}\Gamma(s)^2 \tilde{g}(s; T, G) ds \quad (\operatorname{Re}\xi \geq 0) \quad (5.1.5)$$

with

$$\tilde{g}(s; T, G) = \int_0^\infty y^{s-1}(1+y)^{-\frac{1}{2}+iT} \exp(-\tfrac{1}{4}(G\log(1+y))^2) dy$$

$$(\operatorname{Re}s > 0). \quad (5.1.6)$$

Also we have, by (4.6.5)–(4.6.9),

$$\Xi(\xi; T, G) = \int_0^\infty R(y, \xi) y^{-\frac{1}{2}+\xi}(1+y)^{-\frac{1}{2}+iT} \exp(-\tfrac{1}{4}(G\log(1+y))^2) dy$$

$$(\operatorname{Re}\xi \geq 0) \quad (5.1.7)$$

with

$$R(y, \xi) = \int_0^1 (x(1-x))^{-\frac{1}{2}+\xi}(1+yx)^{-\frac{1}{2}-\xi} dx. \quad (5.1.8)$$

We are going to estimate $\Xi(\xi; T, G)$ asymptotically in various ways. Because of the observation made at (4.5.1) we know that it is of rapid decay with respect to ξ. But what we need now is a result which is uniform in the three parameters ξ, T, G; and this is a somewhat involved task.

To begin with, we introduce a harmless restriction: We assume that T is sufficiently large, and

$$T^a < G < T(\log T)^{-1}, \quad (5.1.9)$$

where a is an arbitrary fixed small positive constant. Although a more stringent assumption is to be introduced in the sequel, we shall show that already on the assumption of (5.1.9) the contribution of the holomorphic cusp-forms is negligible, confirming what we said at the beginning of Section 2.2.

Lemma 5.1 *Let $C > 0$ be an arbitrary fixed constant. Then we have, uniformly for integral $k \geq 1$,*

$$\Xi(2k - \tfrac{1}{2}; T, G) \ll T^{-2k} + (k^2 G)^{-C}, \quad (5.1.10)$$

where the implied constant depends only on C. In particular we have

$$I_{2,h}(T, G) \ll T^{-2} \quad (5.1.11)$$

uniformly for those T, G satisfying (5.1.9).

Proof We may assume that C is an integer. Then we have, by (5.1.5),

$$\Xi(2k-\tfrac{1}{2};T,G)=\sum_{v=0}^{C-2k}(-1)^v\frac{\Gamma(2k+v)^2}{\Gamma(4k+v)\Gamma(v+1)}\tilde{g}(2k+v;T,G)$$
$$+\frac{1}{2\pi i}\int_{(C+\frac{1}{2})}\frac{\Gamma(2k-s)}{\Gamma(2k+s)}\Gamma(s)^2\tilde{g}(s;T,G)ds. \qquad (5.1.12)$$

We have, by (4.1.11) with the present choice of g,

$$\tilde{g}(2k+v;T,G)=(-1)^v\Gamma(2k+v)(\sqrt{\pi}G)^{-1}\int_{-\infty+Di}^{\infty+Di}e^{-(t/G)^2}$$
$$\times\left\{(\tfrac{1}{2}+i(T+t))(\tfrac{3}{2}+i(T+t))\cdots(2k+v+\tfrac{1}{2}+i(T+t))\right\}^{-1}dt$$

for any $D>2k+v$. Moving the path to the real axis we get

$$\tilde{g}(2k+v;T,G)\ll T^{-2k-v}$$

provided (5.1.9) holds, where the implied constant depends only on k, v. On the other hand, to estimate the integral in (5.1.12) we note that we have, by (5.1.6),

$$|\tilde{g}(s;T,G)|\le\tilde{g}(\mathrm{Re}\,s;0,G)\ll G^{-\mathrm{Re}\,s}, \qquad (5.1.13)$$

provided $\mathrm{Re}\,s$ is positive and bounded. This implies that we have, uniformly in $k\ge 1$,

$$\int_{(C+\frac{1}{2})}\frac{\Gamma(2k-s)}{\Gamma(2k+s)}\Gamma(s)^2\tilde{g}(s;T,G)ds\ll(k^2G)^{-C}$$

with the implied constant depending only on C, which proves (5.1.10). The assertion (5.1.11) is a result of combining (5.1.10) with (3.2.22) and the second inequality in (4.4.5). This ends the proof of the lemma. It should be remarked that (2.2.4) implies that we have in fact $I_{2,h}(T,G)\ll T^{-6}$.

We then move to the estimation of $\Xi(ir;T,G)$ with a real r. We shall consider five cases separately according as

(1) $|r|\ge T\log^2 T$,
(2) $2G^{-1}T\log^5 T\le|r|\le T\log^2 T$,
(3) $T^{\frac{1}{2}-a}\le r\le 2G^{-1}T\log^5 T$,
(4) $|r|\le T^{\frac{1}{2}-a}$,
(5) $-2G^{-1}T\log^5 T\le r\le -T^{\frac{1}{2}-a}$,

where a is as in (5.1.9). Naturally some of these conditions are impossible unless we suppose, instead of (5.1.9),

$$T^a<G<2T^{\frac{1}{2}+a}\log^5 T. \qquad (5.1.14)$$

In the discussion below we shall assume this; but obviously we may drop it at any time, for if it is not applicable we need only consider the cases (1), (2), (4). In the first three cases we shall show that $\Xi(ir;T,G)$ is negligibly small; on the other hand in the last two cases we shall get asymptotic expansions.

Lemma 5.2 *In the cases* (1)–(3) *we have*

$$\Xi(ir;T,G) \ll ((1+|r|)T)^{-C} \tag{5.1.15}$$

for any fixed constant $C>0$.

Proof We begin with the case (1). For this purpose we note that we have, more precisely than (5.1.13),

$$\tilde{g}(s;T,G) \ll G^{-\mathrm{Re}\,s}\exp(-|\mathrm{Im}\,s|/T) \tag{5.1.16}$$

with the same condition on $\mathrm{Re}\,s$. To show this we turn the line of integration in (5.1.6) through an angle θ, and observe that we have, putting $y=\lambda e^{i\theta}$ ($\lambda\ge 0$),

$$1+y=(1+2\lambda\cos\theta+\lambda^2)^{\frac{1}{2}}e^{i\vartheta}$$

with

$$|\vartheta| = \arctan\left(\frac{\lambda\sin|\theta|}{1+\lambda\cos\theta}\right) \le \min\{\,|\theta|,\,\lambda\sin|\theta|\,\};$$

and so we have, for all $\lambda\ge 0$,

$$\begin{aligned}\mathrm{Re}\,\{\log^2(1+y)\} &= \tfrac{1}{4}(\log(1+2\lambda\cos\theta+\lambda^2))^2-\vartheta^2\\ &\ge \tfrac{1}{2}(\log(1+\lambda))^2\end{aligned} \tag{5.1.17}$$

provided that θ is small. For the proof of (5.1.16) we need only put $\theta=\mathrm{sgn}(\mathrm{Im}\,s)T^{-1}$ in these and take the absolute value of the integrand.

Having (5.1.16) we shift the path in (5.1.5) with $\xi = ir$ to $\mathrm{Re}\,s=m$, where m is an arbitrary positive integer. We have

$$\begin{aligned}\Xi(ir;T,G) &= \sum_{\nu=0}^{m-1}(-1)^\nu\frac{\Gamma(\nu+\frac{1}{2}+ir)^2}{\Gamma(\nu+1+2ir)\Gamma(\nu+1)}\tilde{g}(\nu+\tfrac{1}{2}+ir;T,G)\\ &\quad+\Xi_m(ir;T,G),\end{aligned} \tag{5.1.18}$$

where $\Xi_m(ir;T,G)$ is the resulting integral over the line $\mathrm{Re}\,s=m$. The bound (5.1.16) implies that this sum over ν is

$$\ll e^{-|r|/T}((1+|r|)G)^{-\frac{1}{2}}\left(1+(|r|/G)^{m-1}\right);$$

and that

$$\Xi_m(ir;T,G) \ll G^{-m} \int_{(m)} |s|^{2m-1}(|s+ir||s-ir|)^{-m}$$
$$\times \exp\left(-\tfrac{1}{2}\pi(2|s|-|s+ir|+|s-ir|)-|s|/T\right)|ds|$$
$$\ll G^{-m}\{(1+|r|)^{-2m}T^{2m}+(1+|r|)^{m-1}\exp(-|r|/(2T))\}, \tag{5.1.19}$$

where the implied constants depend only on m. Taking m sufficiently large in these, we find that we have, for any fixed $C>0$,

$$\Xi(ir;T,G) \ll |r|^{-C} \quad (|r| \geq T\log^2 T) \tag{5.1.20}$$

provided (5.1.9) holds, where the implied constant depends only on C and a. This finishes the case (1).

We next consider the case (2). This time we use the expression (5.1.7). In it we replace the path by the one consisting of the segment L_1 connecting 0 and y^* and the half line L_2 connecting y^* and $+\infty e^{i\mu}$, where

$$\mu = \mathrm{sgn}(r)(T\log^2 T)^{-1}, \quad y^* = G^{-1}\log T + i(1+G^{-1}\log T)\tan\mu.$$

We remark that on the new path we have $R(y,r) \ll 1$ uniformly for all r under consideration since $|\arg(1+xy)| \leq \mu$ in (5.1.8). Then, on noting that $\arg(1+y)=\mu$ on L_2, we see readily that the contribution of L_2 is $O(\exp(-(\log T)^2))$. On the other hand we have on L_1

$$\mathrm{sgn}(r)\arg(y) = \arctan\{(1+G/\log T)\tan(|\mu|)\} \geq G/(2T\log^3 T).$$

Hence we have

$$\Xi(ir;T,G) \ll \exp\left(-\frac{G|r|}{2T(\log T)^3}\right) + \exp(-(\log T)^2) \quad (|r| \leq T\log^2 T);$$

that is, if (2) holds,

$$\Xi(ir;T,G) \ll \exp(-(\log T)^2), \tag{5.1.21}$$

which is negligible.

To deal with the next case (3) we again use (5.1.7) with (5.1.8), but this time we exchange the order of integration: We have

$$\Xi(ir;T,G) = \int_0^1 (x(1-x))^{-1/2+ir}V(x,r;T,G)dx, \tag{5.1.22}$$

where

$$V(x,r;T,G) = \int_0^\infty y^{-1/2+ir}(1+y)^{-1/2+iT}(1+xy)^{-1/2-ir}$$
$$\times \exp(-(\tfrac{1}{2}G\log(1+y))^2)dy. \tag{5.1.23}$$

We turn the path in the last integral through the angle $r^{-\frac{1}{2}}$, and invoke (5.1.17). We readily get

$$\Xi(ir;T,G) \ll e^{-\sqrt{r}} \quad \left(T^{\frac{1}{2}-a} < r < 2TG^{-1}(\log T)^5\right). \tag{5.1.24}$$

This ends the proof of Lemma 5.2.

Lemma 5.3 *In the case* (4) *we have*

$$\Xi(ir;T,G) = \sum_{v=0}^{m-1}(-1)^v \frac{\Gamma(v+\frac{1}{2}+ir)^2}{\Gamma(v+1+2ir)\Gamma(v+1)}\tilde{g}(v+\tfrac{1}{2}+ir;T,G)$$
$$+ O((1+|r|)^{2m-\frac{3}{2}}T^{-m}), \tag{5.1.25}$$

where the implied constant depends only on m.

Proof This is more involved than its predecessors. We are going to improve the estimate (5.1.19) of $\Xi_m(ir;T,G)$. To this end we transform (5.1.6) a little. We have, for $\operatorname{Re} s = m$,

$$\begin{aligned}
&\tilde{g}(s;T,G)\\
&= (\sqrt{\pi}G)^{-1}\int_0^\infty y^{s-1}(1+y)^{-\frac{1}{2}+iT}\int_{-\infty+mi}^{\infty+mi}(1+y)^{it}e^{-(t/G)^2}dtdy\\
&= (\sqrt{\pi}G)^{-1}\int_0^\infty y^{s-1}(1+y)^{-\frac{1}{2}+iT}\int_{-G\log T+mi}^{G\log T+mi}(1+y)^{it}e^{-(t/G)^2}dtdy\\
&\qquad + O(\exp(-(\log T)^2)).
\end{aligned}$$

Hence we have, for $\operatorname{Re} s = m$,

$$\tilde{g}(s;T,G) = (\sqrt{\pi}G)^{-1}\Gamma(s)\int_{-G\log T+mi}^{G\log T+mi}\frac{\Gamma(\frac{1}{2}-i(T+t)-s)}{\Gamma(\frac{1}{2}-i(T+t))}e^{-(t/G)^2}dt$$
$$+ O(\exp(-(\log T)^2)), \tag{5.1.26}$$

where the implied constant depends only on m. Thus we have

$$\Xi_m(ir;T,G) = \left\{\Xi_{m,1}+\Xi_{m,2}\right\}(ir;T,G),$$

where

$$\Xi_{m,1}(ir;T,G) = \frac{1}{2\pi i\sqrt{\pi}G}\int_{(m)}\int_{-G\log T+mi}^{G\log T+mi} e^{-(t/G)^2}$$
$$\times \frac{\Gamma(s)^3\Gamma(\frac{1}{2}+ir-s)\Gamma(\frac{1}{2}-i(T+t)-s)}{\Gamma(\frac{1}{2}+ir+s)\Gamma(\frac{1}{2}-i(T+t))}dtds$$

and

$$\Xi_{m,2}(ir;T,G) \ll \exp(-(\log T)^2)\int_{(m)}\left|\frac{\Gamma(\frac{1}{2}+ir-s)}{\Gamma(\frac{1}{2}+ir+s)}\right||\Gamma(s)|^2|ds|.$$

Stirling's formula implies that

$$\begin{aligned}\Xi_{m,1}(ir;T,G)&\\ \ll T^{-m}G^{-1}\int_{-\infty}^{\infty}\int_{-G\log T}^{G\log T}&(1+|u|)^{3m-\frac{3}{2}}\{(1+|u+r|)(1+|u-r|)\}^{-m}\\ \times\exp\big(-\tfrac{1}{2}\pi(3|u|+|u-r|&-|u+r|+|u+T+t|-T-t)\\ &\qquad -(t/G)^2\big)dtdu;\end{aligned}$$

and thus we have

$$\Xi_{m,1}(ir;T,G) \ll (1+|r|)^{2m-\frac{3}{2}}T^{-m}.$$

On the other hand it is easy to get

$$\Xi_{m,2}(ir;T,G) \ll (1+|r|)^{m-1}\exp(-(\log T)^2).$$

This ends the treatment of the case (4). We note, in passing, that using (5.1.26) with ν instead of m we see that (5.1.25) implies that

$$\Xi(ir;T,G) = \frac{\Gamma(\frac{1}{2}+ir)^3}{\Gamma(1+2ir)}e^{\frac{1}{4}\pi i-\frac{1}{2}\pi r}T^{-\frac{1}{2}-ir} + O\big(((1+|r|)/T)^{\frac{3}{2}}\big) \qquad (5.1.27)$$

when $|r|$ is relatively small.

Lemma 5.4 *In the case* (5) *we have*

$$\begin{aligned}\Xi(ir;T,G) =&\pi i\Big(\frac{2}{T|r|}\Big)^{\frac{1}{2}}\exp\Big[ir\log\frac{|r|}{4eT}-\Big(\frac{Gr}{2T}\Big)^2\Big]\\ &+O\{T^{-\frac{1}{2}}(|r|^{-1}+G^{-1})e^{-\frac{1}{8}(Gr/T)^2}\},\end{aligned} \qquad (5.1.28)$$

where the implied constant is absolute.

Proof This time we start at the expression (5.1.22) again, but we shall apply the saddle point method to $V(x,r;T,G)$. The saddle point is located at $y=y_0$, which is the unique positive root of the equation

$$r/y + T/(1+y) - rx/(1+xy) = 0.$$

Thus we have

$$y_0 = 2|r|(T-|r|+((T-|r|)^2+4xT|r|)^{1/2})^{-1}, \qquad (5.1.29)$$

which is approximately equal to $|r|/T$. This lies between 0 and $3G^{-1}\log^5 T$. We then move the path in (5.1.23) to the one consisting of two segments S_1, S_2 and a half line S_3:

$$S_1 = \{y = \lambda(1 - \eta\exp(\tfrac{1}{4}\pi i)) : 0 \le \lambda \le y_0\},$$
$$S_2 = \{y = y_0(1 + \xi\exp(\tfrac{1}{4}\pi i)) : -\eta \le \xi \le \eta\},$$
$$S_3 = \{y = \lambda(1 + \eta\exp(\tfrac{1}{4}\pi i)) : y_0 \le \lambda\},$$

where η is a small positive constant. On S_1 the integrand of $V(x,r;T,G)$ is

$$\ll \lambda^{-\frac{1}{2}}\exp\Big\{-|r|\arctan\Big(\frac{\eta}{\sqrt{2}-\eta}\Big) + (T+|r|)\arctan\Big(\frac{\eta y_0}{\sqrt{2}+(\sqrt{2}-\eta)y_0}\Big)\Big\}$$
$$\ll \lambda^{-\frac{1}{2}}\exp(-\tfrac{1}{3}\eta^2|r|)$$

provided η is sufficiently small, since we have

$$-|r|\arctan\Big(\frac{\eta}{\sqrt{2}-\eta}\Big) < -2^{-\frac{1}{2}}|r|\eta(1+\tfrac{1}{2}\eta);$$

and

$$(T+|r|)\arctan\Big(\frac{\eta y_0}{\sqrt{2}+(\sqrt{2}-\eta)y_0}\Big) = 2^{-\frac{1}{2}}y_0T\eta + O(|r|y_0)$$
$$= 2^{-\frac{1}{2}}\eta|r| + O(|r|y_0).$$

Hence S_1 contributes $O(\exp(-\frac{1}{3}\eta^2|r|))$ to $V(x,r;T,G)$ uniformly for $0 \le x \le 1$. We consider next S_3. Here the integrand is

$$\ll \lambda^{-\frac{1}{2}}\exp\Big\{|r|\arctan\Big(\frac{\eta}{\sqrt{2}+\eta}\Big) - T\arctan\Big(\frac{\eta y_0}{\sqrt{2}+(\sqrt{2}+\eta)y_0}\Big) - \tfrac{1}{8}(G\log(1+\lambda))^2\Big\},$$

where we have used (5.1.17) with $\theta = \arctan(\eta/(\sqrt{2}+\eta))$. Then, in just the same way as in the case of S_1 we see that S_3 contributes $O(\exp(-\frac{1}{3}\eta^2|r|))$ to $V(x,r;T,G)$ uniformly for $0 \le x \le 1$.

We now have

$$V(x,r;T,G) = V_0(x,r;T,G) + O(\exp(-\tfrac{1}{3}\eta^2|r|)), \tag{5.1.30}$$

where $V_0(x,r;T,G)$ is the contribution of S_2, and the implied constant does not depend on x. By definition we have

$$V_0(x,r;T,G) = y_0e^{\frac{1}{4}\pi i}\int_{-\eta}^{\eta} f_0(\xi)\exp(if(\xi))d\xi, \tag{5.1.31}$$

where

$$f_0(\xi) = (y(1+y)(1+xy))^{-\frac{1}{2}} \exp(-(\tfrac{1}{2}G\log(1+y))^2),$$
$$f(\xi) = r\log y + T\log(1+y) - r\log(1+xy)$$

with $y = y_0(1+\xi\exp(\frac{1}{4}\pi i))$. We have

$$f(\xi) = f(0) + \tfrac{1}{2}f''(0)\xi^2 + \tfrac{1}{6}f'''(0)\xi^3 + O(|r|\xi^4),$$

where

$$f''(0) = i|r|\left(1 - \frac{T}{|r|}\left(\frac{y_0}{1+y_0}\right)^2 - \left(\frac{xy_0}{1+xy_0}\right)^2\right),$$
$$f'''(0) = O(|r|).$$

These imply in particular that $\exp(if(\xi)) \ll \exp(-\frac{1}{3}\xi^2|r|)$ $(-\eta \le \xi \le \eta)$. Thus we may truncate the integral in (5.1.31) at $\xi = \pm\xi_0$, $\xi_0 = |r|^{-\frac{2}{5}}$, so that uniformly for $0 \le x \le 1$,

$$V_0(x, r; T, G) = V_1(x, r; T, G) + O(\exp(-\tfrac{1}{2}|r|^{\frac{1}{5}})), \tag{5.1.32}$$

where $V_1(x, r; T, G)$ is the part corresponding to $-\xi_0 \le \xi \le \xi_0$, and the constant in the error term is absolute. We then note that if $|\xi| \le \xi_0$ we have

$$\exp(if(\xi)) = \exp(if(0) + \tfrac{1}{2}if''(0)\xi^2)\{1 + \tfrac{1}{6}if'''(0)\xi^3 + O(|r|\xi^4 + |r|^2\xi^6)\} \tag{5.1.33}$$

as well as

$$f_0(\xi) = f_0(0)\left\{1 + \frac{f_0'}{f_0}(0)\xi + O((1+Gy_0)^4\xi^2)\right\}, \tag{5.1.34}$$

where $(f_0'/f_0)(0) \ll 1 + (Gy_0)^2$. Here the assertion on $f_0(\xi)$ may require a proof: We have

$$(\log(1+y))^2 = (\log(1+y_0))^2 + b\xi + O((\xi y_0)^2),$$

where

$$b = 2e^{\frac{1}{4}\pi i}\frac{y_0}{1+y_0}\log(1+y_0).$$

The present assumption on r, together with (5.1.9), gives

$$b\xi \ll y_0^2|r|^{-\frac{2}{5}} \ll T^{-2}|r|^{\frac{8}{5}} \ll (G\log T)^{-2}.$$

Thus

$$\exp(-\tfrac{1}{4}(G\log(1+y))^2)$$
$$= \exp(-\tfrac{1}{4}(G\log(1+y_0))^2)(1 - \tfrac{1}{4}b\xi + O((Gy_0(1+Gy_0)\xi)^2)),$$

which implies (5.1.34). Then by (5.1.33) we have, for $|\xi| \le \xi_0$,

$$f_0(\xi)\exp(if(\xi)) = f_0(0)\exp(if(0) + \tfrac{1}{2}if''(0)\xi^2)$$
$$\times \Big\{1 + \frac{f_0'}{f_0}(0)\xi + \tfrac{1}{6}if'''(0)\xi^3 + O((1+(Gy_0)^4)\xi^2$$
$$+ (1+(Gy_0)^2)|r|\xi^4 + |r|^2\xi^6)\Big\}$$

with the implied constant being absolute. This gives

$$V_1(x,r;T,G)$$
$$= e^{\frac{1}{4}\pi i}y_0 f_0(0)\exp(if(0))\Big(\frac{2\pi}{|f''(0)|}\Big)^{\frac{1}{2}}\{1 + O(|r|^{-1}(1+(Gy_0)^4))\}.$$

Thus we have

$$V_1(x,r;T,G)$$
$$= e^{\frac{1}{4}\pi i}\Big(\frac{2\pi}{T}\Big)^{\frac{1}{2}}\exp\Big\{iT\log(1+y_0) + ir\log\frac{y_0}{1+xy_0} - \Big(\frac{Gr}{2T}\Big)^2\Big\}$$
$$\times\Big\{1 + O\Big(\frac{1}{|r|} + \frac{|r|}{T} + \Big(\frac{1}{G} + \frac{G}{T}\Big)\Big(\frac{G}{T}\Big)^3|r|^3\Big)\Big\}.$$

Since

$$T\log(1+y_0) + r\log\frac{y_0}{1+xy_0} = r\log\frac{|r|}{eT} + (x-\tfrac{1}{2})\frac{r^2}{T} + O\Big(\frac{|r|^3}{T^2}\Big),$$

we have the simpler expression

$$V_1(x,r;T,G) = \Big(\frac{2\pi}{T}\Big)^{\frac{1}{2}}\exp\Big\{\tfrac{1}{4}\pi i + ir\log\frac{|r|}{eT} + (x-\tfrac{1}{2})\frac{ir^2}{T} - \Big(\frac{Gr}{2T}\Big)^2\Big\}$$
$$+ O\big(T^{-\frac{1}{2}}(|r|^{-1} + G^{-1})e^{-\frac{1}{8}(Gr/T)^2}\big),$$

which holds uniformly for $0 \le x \le 1$ under the present supposition on G and r. It should be remarked that we have taken account of the effect of the factor $\exp(-(Gr/(2T))^2)$ in this simplification.

Hence we have, by (5.1.22), (5.1.30), and (5.1.32),

$$\Xi(ir;T,G) = \Big(\frac{2\pi}{T}\Big)^{\frac{1}{2}}\exp\Big\{\frac{\pi i}{4} + ir\log\frac{|r|}{eT} - \frac{ir^2}{2T} - \Big(\frac{Gr}{2T}\Big)^2\Big\}\xi(r,T)$$
$$+ O\big(T^{-\frac{1}{2}}(|r|^{-1} + G^{-1})e^{-\frac{1}{8}(Gr/T)^2}\big) \qquad (5.1.35)$$

uniformly for those r belonging to the case (5). Here

$$\xi(r,T) = \int_0^1 (x(1-x))^{-\frac{1}{2}+ir}\exp(ir^2x/T)dx;$$

and our problem has been reduced to the computation of this integral. According to (2.2.8) we have

$$\xi(r,T) = \sqrt{\pi}\exp(-ir\log(r^2/T) + ir^2/T)\Gamma(\tfrac{1}{2}+ir)J_{ir}(r^2/(2T)). \quad (5.1.36)$$

But this expression is rather irrelevant to our present purpose, and we need to appeal to the saddle point method again. The argument is, however, much simpler than the above; so we can be brief: The saddle point is at

$$x_0 = \left\{1 + |r|/(2T) + (1+(r/(2T))^2)^{\frac{1}{2}}\right\}^{-1}, \quad (5.1.37)$$

which is close to $\frac{1}{2}$. We use the path that is the sum of the three segments connecting the points 0, $x_0(1-\eta e^{\frac{1}{4}\pi i})$, $x_0(1+\eta e^{\frac{1}{4}\pi i})$, 1, where η is a small positive constant. We see readily that the first and the third segments contribute $O(\exp(-\frac{1}{2}\eta^2|r|))$ to the integral. In the second part the integration can be restricted to the sub-segment $x = x_0(1+\xi e^{\frac{1}{4}\pi i})$, $-|r|^{-\frac{2}{5}} \le \xi \le |r|^{-\frac{2}{5}}$, within an error $O(\exp(-\frac{1}{2}|r|^{\frac{1}{5}}))$. Thus we have

$$\begin{aligned}\xi(r,T) &= \left(\frac{x_0}{1-x_0}\right)^{\frac{1}{2}}\exp\left(\tfrac{1}{4}\pi i + ir\log(x_0(1-x_0)) + \frac{ir^2}{T}x_0\right)\\ &\times\int_{-|r|^{-2/5}}^{|r|^{-2/5}}\left((1+\xi e^{\frac{1}{4}\pi i})\left(1-\frac{x_0}{1-x_0}\xi e^{\frac{1}{4}\pi i}\right)\right)^{-\frac{1}{2}+ir}\\ &\times\exp\left(\frac{ir^2}{T}x_0\xi e^{\frac{1}{4}\pi i}\right)d\xi + O(\exp(-\tfrac{1}{2}|r|^{\frac{1}{5}})).\end{aligned}$$

After a simplification we see that this integral is equal to

$$\begin{aligned}&\int_{-|r|^{-2/5}}^{|r|^{-2/5}}\exp\left[-\tfrac{1}{2}|r|\left(1+\left(\frac{x_0}{1-x_0}\right)^2\right)\xi^2\right]\\ &\qquad\times\left(1 + c_1\xi + c_2r\xi^3 + O(\xi^2+|r|\xi^4+|r|^2\xi^6)\right)d\xi,\end{aligned}$$

where c_1, c_2 are independent of ξ and bounded absolutely; and the implied constant is also absolute. This gives

$$\xi(r,T) = \left(\frac{2\pi}{|r|}\right)^{\frac{1}{2}}\frac{(x_0(1-x_0))^{\frac{1}{2}+ir}}{(x_0^2+(1-x_0)^2)^{\frac{1}{2}}}\exp\left(\tfrac{1}{4}\pi i + ir^2x_0/T\right) + O(|r|^{-1}) \quad (5.1.38)$$

uniformly for all $r < 0$ and $T > 0$.

We insert the last estimate into (5.1.35) and simplify the result slightly on noting that (5.1.37) implies $x_0 = \frac{1}{2}(1-|r|/(4T)) + O((|r|/T)^3)$. Then we find that uniformly for those r in the case (5)

$$\Xi(ir;T,G)=\pi i\left(\frac{2}{T|r|}\right)^{\frac{1}{2}}(x_0(1-x_0))^{ir}\exp\left[ir\log\frac{|r|}{eT}+i(x_0-\tfrac{1}{2})\frac{r^2}{T}-\left(\frac{Gr}{2T}\right)^2\right]$$
$$+O\left(T^{-\frac{1}{2}}(|r|^{-1}+G^{-1})e^{-\frac{1}{8}(Gr/T)^2}\right)$$

provided (5.1.9) holds. This gives rise to (5.1.28), and we end the discussion of all cases.

Now in order to simplify the situation we introduce the condition

$$T^{\frac{1}{3}}\log^2 T\le G\le T(\log T)^{-1}. \tag{5.1.39}$$

Then the assertions in the last three lemmas can be summarized as follows: If $|r|\le(\log T)^2$, then

$$\Xi(ir;T,G)=\frac{\Gamma(\frac{1}{2}+ir)^3}{\Gamma(1+2ir)}e^{\frac{1}{4}\pi i-\frac{1}{2}\pi r}T^{-\frac{1}{2}-ir}+O\left(((1+|r|)/T)^{\frac{3}{2}}\right); \tag{5.1.40}$$

if $-TG^{-1}\log T\le r\le-(\log T)^2$, then

$$\Xi(ir;T,G)=\pi i\left(\frac{2}{T|r|}\right)^{\frac{1}{2}}\exp\left[ir\log\frac{|r|}{4eT}-\left(\frac{Gr}{2T}\right)^2\right]$$
$$+O\left(T^{-\frac{1}{2}}(|r|^{-1}+G^{-1})e^{-\frac{1}{8}(Gr/T)^2}\right); \tag{5.1.41}$$

and otherwise

$$\Xi(ir;T,G)\ll((1+|r|)T)^{-C} \tag{5.1.42}$$

for any fixed constant $C>0$. Thc implied constants in (5.1.40) and (5.1.41) are absolute; and the one in (5.1.42) depends only on C.

These estimates yield the local explicit formula for the fourth power moment of the zeta-function:

Theorem 5.1 *Let D be an arbitrary positive constant, and let us assume that*

$$T^{\frac{1}{2}}(\log T)^{-D}\le G\le T(\log T)^{-1}. \tag{5.1.43}$$

Then we have

$$\frac{1}{\sqrt{\pi}G}\int_{-\infty}^{\infty}|\zeta(\tfrac{1}{2}+i(T+t))|^4\exp(-(t/G)^2)dt$$
$$=\frac{\pi}{\sqrt{2T}}\sum_{j=1}^{\infty}\alpha_jH_j(\tfrac{1}{2})^3\kappa_j^{-\frac{1}{2}}\sin\left(\kappa_j\log\frac{\kappa_j}{4eT}\right)\exp(-\tfrac{1}{4}(G\kappa_j/T)^2)$$
$$+O((\log T)^{3D+9}), \tag{5.1.44}$$

where the implied constant depends only on D.

Proof We return to (5.1.1), but we suppose (5.1.39). We have

$$I_{2,r}(T,G) \ll (\log T)^4,$$

which is immediate. The part $I_{2,h}(T,G)$ has already been found to be negligible as in (5.1.11). On the other hand, (5.1.3) and (5.1.40)–(5.1.42) give

$$I_{2,c}(T,G) \ll T^{-\frac{1}{2}} \int_{-\infty}^{\infty} \frac{|\zeta(\frac{1}{2}+ir)|^6}{|\zeta(1+2ir)|^2}(1+|r|^{\frac{1}{2}})^{-1} \exp(-\tfrac{1}{5}(Gr/T)^2)dr;$$

and by the classical bounds we have

$$I_{2,c}(T,G) \ll T^{\frac{1}{3}} G^{-\frac{5}{6}} (\log T)^9, \tag{5.1.45}$$

which is negligible. As to the contribution of the discrete spectrum we see that the combination of (3.2.5), (4.4.5), (5.1.3), (5.1.40), and (5.1.42) yields

$$I_{2,d}(T,G) = \frac{1}{2} \sum_{(\log T)^2 \le \kappa_j \le TG^{-1}\log T} \alpha_j H_j(\tfrac{1}{2})^3 \mathrm{Re}\left[\Xi(-i\kappa_j;T,G)\right] + O(T^{-\frac{1}{3}}).$$

Into this we insert the expression (5.1.41), the error term of which contributes in total

$$\ll T^{-\frac{1}{2}} \sum_{j=1}^{\infty} \alpha_j |H_j(\tfrac{1}{2})|^3 \{\kappa_j^{-1} + G^{-1}\} \exp(-\tfrac{1}{8}(\kappa_j G/T)^2)$$
$$\ll T^{\frac{1}{2}} G^{-1}(1+TG^{-2})(\log T)^9,$$

where we have used the mean-value estimates (3.4.4) and (3.5.18). This ends the proof of the theorem.

It would be worth remarking that given the condition (5.1.43) the contribution of the continuous spectrum is infinitesimally small. This is in accordance with what we stated about the nature of $\mathscr{Z}_{2,c}(g)$ in Section 4.8. The lower bound of G in (5.1.43) appears to be essentially optimal in so far as one follows our line of reasoning. Looking at the summands on the right side of (5.1.44) we notice that the factor $T^{-\frac{1}{2}\pm i\kappa_j}$ is actually present. Since this corresponds to a residue of a simple expression involving the logarithmic derivative of the Selberg zeta-function for the full modular group, (5.1.44) suggests the existence of a certain analytic function involving the Riemann zeta-function which has singularities precisely at the zeros of the Selberg zeta-function. These are to be elaborated in Section 5.3.

5.2 Global bounds

Now we move to a more classical object: In this section we shall investigate the size of the remainder term in the asymptotic formula for the unweighted fourth power moment

$$M_2(T) = \int_0^T |\zeta(\tfrac{1}{2} + it)|^4 dt. \tag{5.2.1}$$

We shall show that the main term is of the form $TP_4(\log T)$ with a certain polynomial P_4 of fourth degree. Thus we may put

$$M_2(T) = TP_4(\log T) + E_2(T), \tag{5.2.2}$$

where $E_2(T)$ stands for the error term. We shall give upper bounds for both the individual values and the mean square of $E_2(T)$ as applications of our explicit formula for $\mathscr{Z}_2(g)$.

To this end we introduce two large parameters G and U such that

$$U^{\frac{1}{3}}(\log U)^2 \le G \le \tfrac{1}{2}U(\log U)^{-1}; \tag{5.2.3}$$

and we put

$$U_1 = U - G\log U, \quad U_2 = U + G\log U,$$
$$U_3 = 2U - G\log U, \quad U_4 = 2U + G\log U.$$

Further we put

$$J_2^-(U,G) = \int_{U_2}^{U_3} I_2(T,G)dT; \quad J_2^+(U,G) = \int_{U_1}^{U_4} I_2(T,G)dT.$$

Then we readily get

$$J_2^-(U,G) - \exp(-\tfrac{1}{2}(\log U)^2) \le M_2(2U) - M_2(U)$$
$$\le J_2^+(U,G) + \exp(-\tfrac{1}{2}(\log U)^2). \tag{5.2.4}$$

This reduces our problem (5.2.1) to the estimation of $J_2^\pm$. Since they are analogous, we shall consider only J_2^+ in detail; and for the sake of notational simplicity we shall write J_2 in place of J_2^+ in the sequel (there will be no confusion with the notation for Bessel functions). The decomposition (5.1.1) of $I_2(T,G)$ entails

$$J_2(U,G) = \left\{J_{2,r} + J_{2,d} + J_{2,c} + J_{2,h}\right\}(U,G) \tag{5.2.5}$$

with the obvious arrangement of the terms.

By definition we have

$$J_{2,r}(U,G) = \sum_{\substack{a,b,k,l\geq 0\\ ak+bl\leq 4}} \varpi(a,k;b,l) \int_{-\infty}^{\infty} \mathrm{Re}\left[\left(\frac{\Gamma^{(a)}}{\Gamma}\right)^k \left(\frac{\Gamma^{(b)}}{\Gamma}\right)^l (\tfrac{1}{2}+it)\right]$$

$$\times \frac{1}{\sqrt{\pi}G}\int_{U_1-t}^{U_4-t} e^{-(r/G)^2} drdt + O(\exp(-(\log U)^2)).$$

Stirling's formula gives an asymptotic expansion of the element in the square brackets, and we find easily that

$$J_{2,r}(U,G) = \int_{U}^{2U} p_4(\log t)dt + O(G(\log U)^5), \tag{5.2.6}$$

where p_4 is a certain polynomial of fourth degree whose coefficients are absolute constants. The estimation of $J_{2,h}(U,G)$ is fairly simple; the assertion (5.1.11) gives

$$J_{2,h}(U,G) \ll U^{-1}. \tag{5.2.7}$$

On the other hand, to deal with $J_{2,c}(U,G)$ and $J_{2,d}(U,G)$ we have to study first

$$X(r;P,Q;G) = \int_{P}^{Q} \Xi(ir;T,G)dT.$$

Lemma 5.5 *On the conditions $P,Q \approx U$ and (5.2.3) we have the following estimates: If $|r| \leq (\log U)^2$, then*

$$X(r;P,Q;G) = -\frac{\Gamma(\frac{1}{2}+ir)^2\Gamma(-\frac{1}{2}+ir)}{\Gamma(1+2ir)} e^{\frac{1}{4}\pi i-\frac{1}{2}\pi r}(Q^{\frac{1}{2}-ir} - P^{\frac{1}{2}-ir})$$

$$+O\big(((1+|r|)/U)^{\frac{1}{2}}\big); \tag{5.2.8}$$

if $-UG^{-1}\log U \leq r \leq -(\log U)^2$, then

$$X(r;P,Q;G) = \Xi^*(ir;Q,G) - \Xi^*(ir;P,G)$$

$$+O\big(U^{\frac{1}{2}}|r|^{-1}(|r|^{-1}+G^{-1})\exp(-\tfrac{1}{8}(Gr/P)^2)\big), \tag{5.2.9}$$

where

$$\Xi^*(ir;T,G) = \pi(2T)^{\frac{1}{2}}|r|^{-\frac{3}{2}} \exp\left[ir\log\frac{|r|}{4eT} - \left(\frac{Gr}{2T}\right)^2\right]; \tag{5.2.10}$$

and otherwise

$$X(r;P,Q;G) \ll ((1+|r|)U)^{-C} \tag{5.2.11}$$

for any fixed constant $C>0$. The implied constants in (5.2.8)–(5.2.9) are absolute, and the one in (5.2.11) depends only on C.

Proof The assertions (5.2.8) and (5.2.11) are immediate consequences of (5.1.25) and (5.1.42), respectively. Thus we need to consider (5.2.9) only. We note that then we can assume that $|r|$ is larger than a positive power of U. For the estimate (5.2.8) actually holds for $|r| \leq U^{\frac{1}{2}-a}$ with any fixed $a > 0$ as (5.1.25) shows; and (5.2.9) is a consequence of (5.2.8) when $|r|$ is relatively small.

We integrate the expressions (5.1.22) and (5.1.23) with respect to the variable T. Following the argument of the preceding section up to (5.1.32), we easily get

$$X(r;P,Q;G) = \{X^*(r;Q,G) - X^*(r;P,G)\} + O(\exp(-(\log U)^2))$$

with

$$X^*(r;T,G) = \int_0^1 (x(1-x))^{-1/2+ir} V^*(x,r;T,G)dx.$$

Here

$$V^*(x,r;T,G) = -iy_0 e^{\frac{1}{4}\pi i} \int_{-|r|^{-2/5}}^{|r|^{-2/5}} y^{-1/2+ir}(1+y)^{-1/2+iT}(\log(1+y))^{-1} \times (1+xy)^{-1/2-ir}\exp(-(\tfrac{1}{2}G\log(1+y))^2)d\xi,$$

where $y = y_0(1+\xi e^{\frac{1}{4}\pi i})$ and y_0 is the same as in (5.1.29). With a minor modification the approximation process developed in between (5.1.33) and (5.1.35) gives

$$V^*(x,r;T,G) = e^{-\frac{1}{4}\pi i}\frac{y_0}{\log(1+y_0)} f_0(0)\exp(if(0))\Big(\frac{2\pi}{|f''(0)|}\Big)^{\frac{1}{2}} \times \{1 + O(|r|^{-1}(1+(Gy_0)^4))\},$$

where f_0 and f are the same as in (5.1.31). Hence we have

$$\begin{aligned} &X^*(r;T,G) \\ &= (2\pi T)^{\frac{1}{2}}|r|^{-1}\exp\Big\{-\tfrac{1}{4}\pi i + ir\log\frac{|r|}{eT} - \frac{ir^2}{2T} - \Big(\frac{Gr}{2T}\Big)^2\Big\}\xi(r,T) \\ &+ O\big(T^{\frac{1}{2}}|r|^{-1}(|r|^{-1}+G^{-1})\exp(-\tfrac{1}{8}(Gr/T)^2)\big), \end{aligned}$$

where $\xi(r,T)$ is as in (5.1.35). This and (5.1.38) obviously end the proof of (5.2.9).

Returning to our original problem (5.2.1), we have

Lemma 5.6 *Let us assume that*

$$U^{\frac{1}{2}}(\log U)^4 \leq G \leq U(\log U)^{-1}; \tag{5.2.12}$$

and let M_2, U_j $(1 \le j \le 4)$, p_4 *be as above. Further let us put*

$$J_{2,d}^*(T,G) = \frac{1}{2}\sum_{j=1}^{\infty} \alpha_j H_j(\tfrac{1}{2})^3 \mathrm{Re}\left[\Xi^*(i\kappa_j;T,G)\right]. \tag{5.2.13}$$

Then we have

$$\begin{aligned} J_{2,d}^*(U_3,G) &- J_{2,d}^*(U_2,G) - G(\log U)^6 \\ &\le M_2(2U) - M_2(U) - \int_U^{2U} p_4(\log t)dt \\ &\qquad \le J_{2,d}^*(U_4,G) - J_{2,d}^*(U_1,G) + G(\log U)^6 \end{aligned} \tag{5.2.14}$$

provided U *is sufficiently large.*

Proof The previous lemma gives

$$J_{2,c}(U,G) \ll U^{\frac{1}{2}} \int_{-\infty}^{\infty} \frac{|\zeta(\frac{1}{2}+ir)|^6}{|\zeta(1+2ir)|^2}\{(1+|r|)^{-\frac{3}{2}} + G^{-1}(1+|r|)^{-1}\} \times \exp(-\tfrac{1}{32}(Gr/U)^2)dr;$$

and by the classical bounds we have

$$J_{2,c}(U,G) \ll U^{\frac{1}{2}} + U^{\frac{5}{6}}G^{-\frac{4}{3}}(\log U)^9. \tag{5.2.15}$$

As to $J_{2,d}$ we have, by the last lemma coupled with (3.4.4) and (3.5.18),

$$J_{2,d}(U,G) = J_{2,d}^*(U_4,G) - J_{2,d}^*(U_1,G) + O\big(U^{\frac{1}{2}}(1+UG^{-2})(\log U)^9\big). \tag{5.2.16}$$

Combining (5.2.4)–(5.2.7), (5.2.15), and (5.2.16) we end the proof of the lemma.

As an immediate consequence of (5.2.14) we have

Theorem 5.2 *Let* p_4 *be as above. Then the main term of* $M_2(T)$ *is equal to*

$$\int_0^T p_4(\log t)dt = TP_4(\log T)$$

with a polynomial P_4 *of fourth degree. With this choice of* P_4 *in* (5.2.2) *we have*

$$E_2(T) \ll T^{\frac{2}{3}}(\log T)^8. \tag{5.2.17}$$

Proof In fact, using (3.4.4) and (3.5.18) again, we have

$$J_{2,d}^*(U_j,G) \ll UG^{-\frac{1}{2}}(\log U)^9.$$

Inserting this into (5.2.14) and choosing G optimally, we get

$$E_2(2U) - E_2(U) \ll U^{\frac{2}{3}}(\log U)^8,$$

which gives the theorem.

Next, we shall refine (5.2.17). To this end we consider the set of points $t \in [T, 2T]$ such that $|E_2(t)| \geq V$ ($V \ll T^{\frac{2}{3}}(\log T)^8$). Naturally it is expected that the set becomes thinner as V increases. In order to confirm this in a quantitative formulation, we observe first that if $E_2(t)$ is large at a point then it continues to be large in an interval of considerable length. More precisely, let us assume that $\pm E_2(t) \geq V$. Since $\pm M_2(T \pm u) \geq M_2(T)$ for any $u \geq 0$, we have

$$V \leq \pm E_2(T) \leq \pm E_2(T \pm u) + O(u(\log T)^4)$$

provided u is relatively small compared with T. Thus we have, for all $0 \leq u \leq V(\log T)^{-5}$,

$$\pm E_2(T \pm u) \geq \tfrac{1}{2}V.$$

This means that the middle points of the intervals throughout which $|E_2(t)| \geq \frac{1}{2}V$ are at least $V(\log T)^{-5}$ well-spaced.

In view of the above observation we now assume that R points t_r are such that

$$T \leq t_1 < \ldots < t_R \leq 2T, \quad t_{r+1} - t_r \geq V(\log T)^{-8}$$
$$(0 \leq r \leq R-1); \quad |E_2(t_r)| \geq V,$$

where

$$T^{\frac{1}{2}}(\log T)^{11} \leq V \leq T^{\frac{2}{3}}(\log T)^9.$$

We are going to bound R in terms of T and V. For this purpose we assume further that we have $E_2(t_r) \geq V$ for all t_r's. Obviously this causes no loss of generality. Then, in (5.2.14) we put

$$U = 2^{-l}t_r \quad (l = 1, 2, \ldots, L), \quad G = V(\log T)^{-7},$$

where L is to satisfy

$$2^{-L}T \approx T^{\frac{3}{4}}.$$

We note that by (5.2.17) we have

$$E_2(t_r) = \sum_{l=1}^{L} \{E_2(2^{1-l}t_r) - E_2(2^{-l}t_r)\} + O(T^{\frac{1}{2}}(\log T)^8);$$

and

$$\frac{1}{2}RV \leq \sum_{l=1}^{L}\sum_{r=1}^{R}\{E_2(2^{1-l}t_r) - E_2(2^{-l}t_r)\}.$$

Thus by (5.2.14) we have

$$\frac{1}{4}RV \leq \sum_{l=1}^{L}\sum_{r=1}^{R}\{J^*_{2,d}(\tau(r,l),G) - J^*_{2,d}(\tau'(r,l),G)\}, \tag{5.2.18}$$

where $\tau(r,l) = 2^{1-l}t_r + G\log T$ and $\tau'(r,l) = 2^{-l}t_r - G\log T$. We have, by (5.2.10),

$$\sum_{r=1}^{R} J^*_{2,d}(\tau(r,l);G)$$
$$\ll \sum_{j=1}^{\infty} \alpha_j\kappa_j^{-\frac{3}{2}}|H_j(\tfrac{1}{2})|^3\left|\sum_{r=1}^{R}\tau(r,l)^{\frac{1}{2}+i\kappa_j}\exp(-(G\kappa_j/(2\tau(r,l)))^2)\right|.$$

This sum over j can be truncated so that $\kappa_j \leq TG^{-1}\log T$; and then we replace the exponentiated factor by

$$\frac{1}{2\pi i}\int_{(\beta)}\Gamma(z)\big(G\kappa_j/(2\tau(r,l))\big)^{-2z}dz,$$

where $\beta = (\log T)^{-1}$. We have

$$\sum_{r=1}^{R} J^*_{2,d}(\tau(r,l);G) \ll 1 + (\log T)\max_K K^{-\frac{3}{2}}\int_{(\beta)} W(K,l;z)|\Gamma(z)||dz|, \tag{5.2.19}$$

where K runs over the numbers $2^{-\nu}TG^{-1}\log T$ $(\nu = 1,2,\ldots)$, and

$$W(K,l;z) = \sum_{K\leq\kappa_j\leq 2K}\alpha_j|H_j(\tfrac{1}{2})|^3\left|\sum_{r=1}^{R}\tau(r,l)^{\frac{1}{2}+2z+i\kappa_j}\right|.$$

By Cauchy's inequality and the mean value (3.5.18), we have

$$W(K,l;z)^2 \ll K^2(\log K)^{15}\sum_{K\leq\kappa_j\leq 2K}\alpha_j|H_j(\tfrac{1}{2})|^2\left|\sum_{r=1}^{R}\tau(r,l)^{\frac{1}{2}+2z+i\kappa_j}\right|^2.$$

To this sum over κ_j we may apply (3.4.3). Since

$$\tau(r+1,l) - \tau(r,l) = 2^{1-l}(t_{r+1} - t_r) \geq 2^{1-l}V(\log T)^{-5},$$

we have, uniformly in z, $\operatorname{Re} z = \beta$,

$$W(K,l;z)^2 \ll K^3(\log K)^{16}(K + TV^{-1}(\log T)^5)RT2^{-l}.$$

Inserting this into (5.2.19) we get

$$\sum_{r=1}^{R} J_{2,d}^{*}(\tau(r,l);G) \ll 2^{-l/2} R^{\frac{1}{2}} T V^{-\frac{1}{2}} (\log T)^{14}.$$

Obviously the same holds if we replace $\tau(r,l)$ by $\tau'(r,l)$ on the left side. Thus by (5.2.18) we get

$$RV \ll R^{\frac{1}{2}} T V^{-\frac{1}{2}} (\log T)^{14},$$

or

$$R \ll T^2 V^{-3} (\log T)^{28}. \tag{5.2.20}$$

This ends our discussion on the large values of $E_2(T)$.

We are now able to prove

Theorem 5.3 *We have*

$$\int_0^T E_2(t)^2 dt \ll T^2 (\log T)^{22}. \tag{5.2.21}$$

Proof Since the part where $|E_2(t)| \le T^{\frac{1}{2}}(\log T)^{11}$ is trivial, we may restrict the integration to the set $\mathcal{S}$ where $|E_2(t)| \ge T^{\frac{1}{2}}(\log T)^{11}$. Let us consider the subset $\mathcal{S}_V$ where $V \le |E_2(t)| < 2V$, $t \in \mathcal{S} \cap [T/2, T]$. We divide the interval $[T/2, T]$ into subintervals of length $V(\log T)^{-8}$; the end subintervals can be shorter than V. Then the number of those subintervals which contain a point of $\mathcal{S}_V$ is bounded by (5.2.20). Hence we have

$$\int_{\mathcal{S}_V} E_2(t)^2 dt \ll T^2 (\log T)^{20},$$

which obviously ends the proof.

5.3 Eigen-peaks

The aim of this section is to indicate that the mean square estimate (5.2.21) is close to the best possible. That is, we are going to show that both $E_2(T)$ and $-E_2(T)$ are often greater than a constant multiple of $T^{\frac{1}{2}}$. This will be achieved via a meromorphic continuation to the domain $\operatorname{Re}\omega < 1$ of the function

$$Z_2(\omega) = \int_1^\infty |\zeta(\tfrac{1}{2} + it)|^4 t^{-\omega} dt, \tag{5.3.1}$$

which is clearly regular for $\operatorname{Re}\omega > 1$. The continuation is a consequence of the explicit formula for $\mathscr{Z}_2(g)$. It will turn out in particular that there are infinitely many simple poles on the straight line $\operatorname{Re}\omega = \frac{1}{2}$, and they are of the form

$$\tfrac{1}{2} + i\kappa$$

with $\kappa^2 + \frac{1}{4}$ belonging to the discrete spectrum of the non-Euclidean Laplacian over the full modular group. From this fact the above assertion on $E_2(T)$ follows immediately with the aid of a function-theoretic argument. We shall give a detailed proof of the continuation of $Z_2(\omega)$ to the region $\operatorname{Re}\omega > 0$; further continuation may be left out, since it is just a matter of technicality.

To begin with, we introduce the function

$$Y(\omega, D) = \frac{1}{2}\int_{-\infty}^{\infty} |\zeta(\tfrac{1}{2} + it)|^4 (t^2 + D^2)^{-\omega/2} dt,$$

where $D > 0$ is an arbitrary constant. We have

$$Z_2(\omega) = Y(\omega, D) + h(\omega, D), \tag{5.3.2}$$

where $h(\omega, D)$ is a function regular for $\operatorname{Re}\omega > 0$. Also we have, integrating by parts,

$$Y(\omega, D) = q_5((\omega-1)^{-1}) + h_1(\omega, D) + \omega \int_0^{\infty} tE_2(t)(t^2+D^2)^{-\omega/2-1} dt, \tag{5.3.3}$$

where q_5 is a polynomial of fifth degree with constant coefficients; and h_1 is regular for $\operatorname{Re}\omega > 0$. Theorem 5.3 implies that the last integral is uniformly convergent for $\operatorname{Re}\omega > \frac{1}{2}$. Thus $Z_2(\omega)$ is regular for $\operatorname{Re}\omega > \frac{1}{2}$ except for the pole at $\omega = 1$ of order five.

Here it would be worth trying to visualize the graph of

$$\left| \int_0^T tE_2(t)(t^2 + D^2)^{-\omega/2-1} dt \right|$$

over the ω-plane. When T increases, peaks corresponding to the eigenvalues should emerge and grow.

Now let us suppose for a moment that ω is real. We put $g(t) = f_\omega(t, D)$ in Theorem 4.2 with

$$f_\omega(t, D) = \frac{1}{2}(t^2 + D^2)^{-\omega/2}.$$

Obviously this satisfies the basic assumption, provided $D, \omega > A$, though the condition on ω will of course be relaxed later. On this condition we have the spectral decomposition

$$Y(\omega, D) = q_5((\omega-1)^{-1}) + h_2(\omega, D) + Y_d(\omega, D) + Y_c(\omega, D) + Y_h(\omega, D), \quad (5.3.4)$$

where h_2 is regular for $\mathrm{Re}\,\omega > 0$; and

$$Y_d(\omega, D) = \sum_{j=1}^{\infty} \alpha_j H_j(\tfrac{1}{2})^3 \Lambda(\kappa_j, \omega; D),$$

$$Y_c(\omega, D) = \pi^{-1} \int_{-\infty}^{\infty} \frac{|\zeta(\frac{1}{2}+ir)|^6}{|\zeta(1+2ir)|^2} \Lambda(r, \omega; D)dr,$$

$$Y_h(\omega, D) = \sum_{k=1}^{\infty} \sum_{j=1}^{\vartheta(2k)} \alpha_{j,2k} H_{j,2k}(\tfrac{1}{2})^3 \Lambda((\tfrac{1}{2}-2k)i, \omega; D).$$

Modifying (4.7.2) slightly, we have

$$\Lambda(r, \omega; D) = \frac{1}{4}\Big(1 + \frac{i}{\sinh(\pi r)}\Big)\Xi(ir, \omega; D) + \frac{1}{4}\Big(1 - \frac{i}{\sinh(\pi r)}\Big)\Xi(-ir, \omega; D)$$

if r is real, and

$$\Lambda((\tfrac{1}{2}-2k)i, \omega; D) = \Xi(2k - \tfrac{1}{2}, \omega; D)$$

if k is a positive integer, where

$$\Xi(\xi, \omega; D) = \frac{\Gamma(\frac{1}{2}+\xi)^2}{\Gamma(1+2\xi)} \int_0^{\infty} y^{\xi-\frac{1}{2}}(1+y)^{-\frac{1}{2}}(f_\omega(\cdot, D))_c(\log(1+y)) \times F(\tfrac{1}{2}+\xi, \tfrac{1}{2}+\xi; 1+2\xi; -y)dy.$$

We remark that the formula (1.1.16) gives

$$(f_\omega(\cdot, D))_c(x) = 2\sqrt{\pi}(x/(2D))^{\frac{1}{2}(\omega-1)} K_{\frac{1}{2}(\omega-1)}(Dx)/\Gamma(\tfrac{1}{2}\omega).$$

Hence, using (3.3.40), we have

$$\Xi(\xi, \omega; D) = \int_0^{\infty} y^{\xi+\omega-\frac{3}{2}} P(y, \xi) Q(y, \omega; D)dy,$$

where

$$P(y, \xi) = \int_0^1 (x(1-x))^{\xi-\frac{1}{2}}(1+yx)^{-\xi-\frac{1}{2}}dx$$

and

$$Q(y, \omega; D) = 2\sqrt{\pi}(2D)^{\frac{1}{2}(1-\omega)} y^{1-\omega}(1+y)^{-\frac{1}{2}} \times (\log(1+y))^{\frac{1}{2}(\omega-1)} K_{\frac{1}{2}(\omega-1)}(D\log(1+y))/\Gamma(\tfrac{1}{2}\omega). \quad (5.3.5)$$

Thus the analytic continuation of $Y(\omega, D)$ depends on the property of $\Xi(\xi, \omega; D)$ as a function of two complex variables ξ and ω:

Lemma 5.7 *Let us assume either that ξ is in a fixed vertical strip contained in the half plane $\operatorname{Re}\xi \geq -\frac{1}{8}$, or that ξ is on the half line $[1, \infty)$. Then, for any bounded ω with $\operatorname{Re}\omega \geq -\frac{1}{8}$, the function*

$$\Xi_1(\xi, \omega; D) = (\xi + \omega - \tfrac{1}{2})\Xi(\xi, \omega; D)$$

is regular and

$$\Xi_1(\xi, \omega; D) \ll |\xi|^{-\frac{1}{4}D}, \tag{5.3.6}$$

provided D is sufficiently large.

Proof We consider first the case where ξ is in a vertical strip; thus we may assume that there is a positive constant c such that

$$-\tfrac{1}{8} \leq \operatorname{Re}\xi \leq c \leq \operatorname{Im}\xi \tag{5.3.7}$$

and

$$-\tfrac{1}{8} \leq \operatorname{Re}\omega \leq c, \quad |\operatorname{Im}\omega| \leq c. \tag{5.3.8}$$

The asymptotic property of the K-Bessel functions implies that $Q(y, \omega; D)$ decays fast when y tends to infinity. On the other hand the behavior of $Q(y, \omega; D)$ when y is close to 0 can be inferred from the definitions (1.1.24)–(1.1.25). More precisely, we have, for each $p \geq 0$,

$$\left(\frac{\partial}{\partial y}\right)^p Q(y, \omega; D) \ll \begin{cases} (1 + y^{1-p-\operatorname{Re}\omega})\log(1/y) & \text{as } y \to 0^+, \\ y^{-\frac{1}{2}D} & \text{as } y \to +\infty. \end{cases} \tag{5.3.9}$$

Also we have

$$\left(\frac{\partial}{\partial y}\right)^p P(y, \xi) \ll |\xi|^p, \tag{5.3.10}$$

uniformly for $y \geq 0$. Thus we see that the expression

$$\Xi_1(\xi, \omega; D) = -\int_0^\infty y^{\xi+\omega-\frac{1}{2}} \frac{\partial}{\partial y}\big[P(y, \xi)Q(y, \omega; D)\big]\,dy \tag{5.3.11}$$

yields an analytic continuation of $\Xi_1(\xi, \omega; D)$ to the domain defined by the conditions (5.3.7) and (5.3.8).

To show the estimate (5.3.6) in the present case, we turn the line of integration in (5.3.11) through a small positive angle θ. We have

$$\Xi_1(\xi,\omega;D) = -\exp((\xi+\omega+\tfrac{1}{2})\theta i) \times \int_0^\infty y^{\xi+\omega-\frac{1}{2}} \Big\{ \frac{\partial}{\partial z}[P(z,\xi)Q(z,\omega;D)] \Big\}_{z=ye^{\theta i}} dy. \quad (5.3.12)$$

The estimate (5.3.9) holds for $(\partial/\partial z)^p Q(z,\omega;D)$ with $z = ye^{\theta i}$ too; and also we have, instead of (5.3.10),

$$\big\{(\partial/\partial z)^p P(z,\xi)\big\}_{z=ye^{\theta i}} \ll |\xi|^p \exp\big((\operatorname{Im}\xi)\arg(1+ye^{\theta i})\big).$$

We then divide the integral in (5.3.12) into two parts corresponding to $0 \le y \le 1$ and the rest; and this implies a decomposition of $\Xi_1(\xi,\omega;D)$. If θ is small, the first part is

$$\begin{aligned} &\ll |\xi| \exp\Big(-\Big(\theta - \arctan\Big(\frac{\sin\theta}{1+\cos\theta}\Big)\Big)\operatorname{Im}\xi\Big) \\ &\ll \exp\big(-\tfrac{1}{3}\theta \operatorname{Im}\xi\big); \end{aligned}$$

and the second part is

$$\begin{aligned} &\ll |\xi| \int_1^\infty y^{-D/3} \exp\Big(-\Big(\theta - \arctan\Big(\frac{y\sin\theta}{1+y\cos\theta}\Big)\Big)\operatorname{Im}\xi\Big)dy \\ &\ll |\xi| \int_1^\infty y^{-D/3} \exp\Big(-\frac{\theta}{2y}\operatorname{Im}\xi\Big)dy \\ &\ll |\xi|^{-D/4}. \end{aligned}$$

This obviously ends the proof of (5.3.6) in the case where (5.3.7) and (5.3.8) hold.

Next, we consider the case where $\xi \ge 1$, and the condition (5.3.8) holds. Here the regularity of $\Xi_1(\xi,\omega;D)$ is easy to check; thus let us prove the decay property (5.3.6) only. To this end we note that we now have, for $y \ge 0$,

$$\begin{aligned} (\partial/\partial y)^p P(y,\xi) &\ll \xi^p \int_0^1 \Big(\frac{x(1-x)}{1+yx}\Big)^{\xi-\frac{1}{2}} dx \\ &\ll \xi^p \big(1+\sqrt{1+y}\big)^{1-2\xi}, \end{aligned}$$

which can be shown by computing the maximum of the integrand. We insert this into (5.3.11). The part corresponding to $0 \le y \le 1$ is easily seen to be $O(2^{-\xi})$. The remaining part is

$$\ll \xi \int_1^\infty y^{-D/3} \Big(\frac{\sqrt{y}}{1+\sqrt{1+y}}\Big)^{2\xi} dy,$$

which is much smaller than $\xi^{-D/2}$. This ends the proof of Lemma 5.7.

We may now state the result on the analytic continuation of $Z_2(\omega)$:

Theorem 5.4 *The function $Z_2(\omega)$ is meromorphic over the entire complex plane. In particular, in the half plane $\omega > 0$ it has a simple pole of order five at $\omega = 1$ and infinitely many simple poles of the form $\frac{1}{2} \pm i\kappa$; all other poles are of the form $\frac{1}{2}\rho$. Here $\kappa^2 + \frac{1}{4}$ is in the discrete spectrum of the non-Euclidean Laplacian over the full modular group, and ρ is a complex zero of the Riemann zeta-function. More precisely the residue at the pole $\omega = \frac{1}{2} + i\kappa$ is equal to*

$$(\tfrac{1}{2}\pi)^{\frac{1}{2}} \left(2^{-i\kappa} \frac{\Gamma(\frac{1}{2}(\frac{1}{2} - i\kappa))}{\Gamma(\frac{1}{2}(\frac{1}{2} + i\kappa))} \right)^3 \Gamma(2i\kappa) \cosh(\pi\kappa) \sum_{\kappa_j = \kappa} \alpha_j H_j(\tfrac{1}{2})^3, \tag{5.3.13}$$

where the summands are as above.

Proof We note first that we have

$$Y_h(\omega, D) = \sum_{k=1}^{\infty} \sum_{j=1}^{\vartheta(2k)} \alpha_{j,2k} H_{j,2k}(\tfrac{1}{2})^3 \Xi_1(2k - \tfrac{1}{2}, \omega; D)/(\omega + 2k - 1).$$

Lemma 5.7 and the bounds (3.2.22), (4.4.5) imply that this sum is regular for $\operatorname{Re}\omega \geq -\frac{1}{8}$. On the other hand we have

$$\begin{aligned} &Y_d(\omega, D) \\ &= \frac{1}{4} \sum_{j=1}^{\infty} \alpha_j H_j(\tfrac{1}{2})^3 \left(1 + \frac{i}{\sinh(\pi\kappa_j)}\right) \Xi_1(i\kappa_j, \omega; D)/(\omega - \tfrac{1}{2} + i\kappa_j) \\ &+ \frac{1}{4} \sum_{j=1}^{\infty} \alpha_j H_j(\tfrac{1}{2})^3 \left(1 - \frac{i}{\sinh(\pi\kappa_j)}\right) \Xi_1(-i\kappa_j, \omega; D)/(\omega - \tfrac{1}{2} - i\kappa_j). \end{aligned}$$

Then Lemma 5.7 and the bounds (3.2.5), (4.4.5) imply that $Y_d(\omega, D)$ is regular for $\operatorname{Re}\omega \geq -\frac{1}{8}$ except for the simple poles at $\omega = \frac{1}{2} \pm \kappa i$, where κ runs over the set of the distinct elements in the set $\{\kappa_j\}$ such that

$$M(\kappa) = \sum_{\kappa_j = \kappa} \alpha_j H_j(\tfrac{1}{2})^3 \neq 0. \tag{5.3.14}$$

According to Theorem 3.2 there are infinitely many κ that satisfy this non-vanishing condition. Thus $Y_d(\omega, D)$ has indeed infinitely many simple poles on the straight line $\operatorname{Re}\omega = \frac{1}{2}$.

Let us compute the residue $R(\kappa)$ of $Y_d(\omega, D)$ at the pole $\omega_\kappa = \frac{1}{2} + i\kappa$.

We have

$$R(\kappa) = \frac{1}{4}M(\kappa)\Big(1 - \frac{i}{\sinh(\pi\kappa)}\Big)\Xi_1(-i\kappa, \omega_\kappa; D).$$

By (5.3.11) we have

$$\begin{aligned}\Xi_1(-i\kappa, \omega_\kappa; D) &= P(0, -i\kappa)Q(0, \omega_\kappa; D)\\ &= \frac{\Gamma(\frac{1}{2} - i\kappa)^2}{\Gamma(1 - 2i\kappa)}Q(0, \omega_\kappa; D).\end{aligned}$$

The formula (5.3.5) gives

$$Q(0, \omega; D) = \frac{\pi^{\frac{3}{2}}2^{1-\omega}}{\Gamma(\frac{1}{2}\omega)\Gamma(\frac{1}{2}(\omega+1))\cos(\frac{1}{2}\pi\omega)}$$

provided $\mathrm{Re}\,\omega < 1$. After a rearrangement we get the assertion (5.3.13).

Next, we consider the contribution of the continuous spectrum. We have

$$\begin{aligned}Y_c(\omega, D) = \frac{1}{2\pi i}\int_{(0)} &\frac{\zeta(\frac{1}{2}+r)^3\zeta(\frac{1}{2}-r)^3}{\zeta(1+2r)\zeta(1-2r)}\\ &\times\Big(1 - \frac{1}{\sin(\pi r)}\Big)\Xi_1(r, \omega; D)(r - \tfrac{1}{2} + \omega)^{-1}dr,\end{aligned}$$

where the path is the straight line $\mathrm{Re}\,r = 0$; note that here it is again assumed that $\mathrm{Re}\,\omega$ is sufficiently large. Let L be an arbitrary positive number such that $\zeta(s) \neq 0$ on the lines $\mathrm{Re}\,s = \pm 2L$, and also $|\mathrm{Im}\,\omega| < L/2$. We then move the path in the last integral to the one that is the result of connecting the points $-i\infty$, $-iL$, $\frac{3}{4} - iL$, $\frac{3}{4} + iL$, iL, $i\infty$ with straight lines. The resulting integral is regular for $\mathrm{Re}\,\omega \geq -\frac{1}{8}$ by virtue of Lemma 5.7. In this procedure we encounter poles at $r = \frac{1}{2}$ and $\frac{1}{2}(1-\rho)$, where ρ runs over complex zeros of the zeta-function such that $|\mathrm{Im}\,\rho| < 2L$. As a function of ω the residue at $r = \frac{1}{2}$ is regular for $\mathrm{Re}\,\omega \geq -\frac{1}{8}$, except for the pole at $\omega = 0$. Further, the residue at $r = \frac{1}{2}(1-\rho)$ is regular for $\mathrm{Re}\,\omega \geq -\frac{1}{8}$, except for the pole at $\omega = \frac{1}{2}\rho$. Since L is arbitrary, this proves that $Y_c(\omega, D)$ admits a meromorphic continuation at least to the domain $\mathrm{Re}\,\omega \geq -\frac{1}{8}$. This ends the proof of the theorem.

Now, returning to the issue stated at the beginning of this section, we have the following $\Omega_\pm$-result:

Theorem 5.5 *There is an explicitly computable constant $C > 0$ such that* *the inequalities*

$$E_2(T) > CT^{\frac{1}{2}}, \quad E_2(T) < -CT^{\frac{1}{2}} \tag{5.3.15}$$

are both attained infinitely often.

Proof Let us assume that

$$E_2(T) + cT^{\frac{1}{2}} > 0 \quad (T > T_0)$$

with a certain $c > 0$. Let $E_2^*(T)$ be equal to $E_2(T) + cT^{\frac{1}{2}}$ for $T > T_0$, and 0 otherwise. Further let us put

$$W(s) = \int_1^\infty E_2^*(T)T^{-s-1}dT. \tag{5.3.16}$$

From (5.3.2), (5.3.3), and Theorem 5.4 the function $W(s)$ is meromorphic in the half plane $\mathrm{Re}\, s > 0$. It has a simple pole at $s = \frac{1}{2}$ with the residue c, and is regular for $\mathrm{Re}\, s > \frac{1}{2}$. Since $E_2^*(T) \geq 0$, a well-known assertion in the theory of functions, i.e., Landau's lemma, implies that the integral in (5.3.16) is absolutely convergent for $\mathrm{Re}\, s > \frac{1}{2}$. In particular, taking the absolute values of both sides of (5.3.16), we have

$$|W(\sigma + it)| \leq W(\sigma) \quad (\sigma > \tfrac{1}{2}).$$

We then let $\kappa = \kappa_0$ satisfy (5.3.14). We obviously have

$$|\lim_{\sigma \to \frac{1}{2}^+} ((\sigma + i\kappa_0) - (\tfrac{1}{2} + i\kappa_0))W(\sigma + i\kappa_0)| \leq \lim_{\sigma \to \frac{1}{2}^+} (\sigma - \tfrac{1}{2})W(\sigma).$$

Thus we get, by (5.3.13),

$$C(\kappa_0) \leq c,$$

where

$$C(\kappa_0) = (\tfrac{1}{2}\pi)^{\frac{1}{2}} |\Gamma(2i\kappa_0)| \cosh(\pi\kappa_0)|M(\kappa_0)|.$$

Hence, if $C < C(\kappa_0)$, then the inequality $E_2(T) < -CT^{\frac{1}{2}}$ should be attained at least once, and consequently infinitely often. The opposite sign case can be treated by considering $-E_2(T)$. This ends the proof of the theorem.

5.4 Notes for Chapter 5, and epilogue

The analyses developed in the first two sections are improved versions of the corresponding parts of Motohashi [47] and Ivić and Motohashi [21].

The third section is taken from Motohashi [53]; a minor error in the concluding remark of [53] has been corrected in (5.3.13). To give a detailed history of the mean-value problem preceding these works is outside the scope of this monograph. Readers are referred to Ivić [20]. Here we shall rather try to develop some asymptotic thoughts extrapolating what has been proved rigorously in the above and elsewhere. But before giving the afterword it should be stressed that Jutila [32] succeeded recently in extending two of our main results, i.e., (5.2.17) and (5.2.21), to the mean square of automorphic L-functions attached to Maass waves which is a proper analogue of the fourth power moment for the zeta-function. Also he showed that an analogue of (5.3.15) holds for those L-functions under a feasible hypothesis. Further, we know already that the same holds for L-functions attached to holomorphic cusp-forms (see [50]). Thus readers should note that what we are dwelling on below has in fact a more global background than what is actually described.

To begin with we shall exhibit the counterparts for the mean square situation of Theorems 5.1–5.4: As a consequence of Theorem 4.1 we have the asymptotic formula

$$I_1(T,G) = 2^{\frac{3}{4}}\pi^{\frac{1}{4}}T^{-\frac{1}{4}}\sum_{n=1}^{\infty}(-1)^n d(n)n^{-\frac{1}{4}}\sin(f(T,n)) \times \exp(-\pi n G^2/(2T)) + O(\log T), \tag{5.4.1}$$

where $T^{\frac{1}{4}} \le G \le T(\log T)^{-1}$ and

$$f(T,n) = 2T\operatorname{arcsinh}((\pi n/(2T))^{\frac{1}{2}}) + (2\pi n T + \pi^2 n^2)^{\frac{1}{2}} - \tfrac{1}{4}\pi.$$

This obviously corresponds to (5.1.44). We next put

$$\int_0^T |\zeta(\tfrac{1}{2}+it)|^2 dt = T(\log T + 2c_E - 1 - \log(2\pi)) + E_1(T).$$

Then the parallels of (5.2.17), (5.2.21), and (5.3.15) are, respectively,

$$E_1(T) \ll T^{\frac{1}{3}}(\log T)^2, \tag{5.4.2}$$

$$\int_0^T E_1(t)^2 dt \ll T^{\frac{3}{2}}, \tag{5.4.3}$$

$$E_1(T) = \Omega_{\pm}(T^{\frac{1}{4}}). \tag{5.4.4}$$

These can be deduced from (4.1.16) in a relatively simple way. It is in fact known (see Ivić [20]) that adding refined arguments one may improve them considerably. Thus the exponent $\frac{1}{3}$ in (5.4.2) can be replaced by a smaller value; the inequality in (5.4.3) by an asymptotic expression that implies (5.4.3) is sharp; and (5.4.4) by $\Omega_{\pm}(f(T)T^{\frac{1}{4}})$ with an $f(T)$ which tends to infinity with T.

Hence the fourth power moment situation looks quite like the mean square situation. There are, however, notable differences between them. For instance, no arguments have been found which can yield any exponents smaller than $\frac{2}{3}$ in (5.2.17). The asymptotic version of (5.2.21) has not been proved yet, although it appears highly plausible that it should hold. Further, any improvements of (5.3.15) that match the above mentioned improvement upon (5.4.3) seem to require very refined information about the distribution of the numbers κ_j that is beyond the present-day technology. Thus the analogy between $\mathscr{Z}_1(g)$ and $\mathscr{Z}_2(g)$ ends here, at least presently.

Nevertheless the sharp resemblance between Theorems 5.1–5.4 and the assertions (5.4.1)–(5.4.4) never appears to be a minor episode, and we may be a bit daring in forming our opinion as we did in the survey article Motohashi [52]. We shall reproduce here the plausible inference developed in [52, Section 4] while adding some new thoughts. This concerns the higher power moments $I_k(T,G)$, $k \geq 3$.

We observe that in (5.4.1) we obviously have

$$\sum_{n\leq x} 1 \ll x, \qquad \sum_{n\leq x} d(n) \ll x\log x,$$

and in (5.1.44), less obviously,

$$\sum_{\kappa_j\leq x} 1 \ll x^2, \qquad \sum_{\kappa_j\leq x} \alpha_j|H_j(\tfrac{1}{2})|^3 \ll x^2(\log x)^8.$$

Here the last follows from (3.4.4) and (3.5.18). The next to the last is a consequence of the spectral expansion (1.1.45) applied to a suitably chosen kernel function, i.e., Selberg's trace formula. Thus it seems reasonable to set forth the following hypothetical formula for $I_3(T,G)$: There are objects ω and a_ω such that

$$\sum_{\omega\leq x} 1 \ll x^3, \qquad \sum_{\omega\leq x} |a_\omega| \ll x^3(\log x)^{c_1},$$

and

$$I_3(T,G) = c_2T^{-\frac{3}{4}}\sum_{\omega} a_\omega\omega^{-\frac{3}{4}}\exp(-c_3(\omega/T)^3G^2) + O((\log T)^{c_4}) \quad (5.4.5)$$

for $T^{\frac{3}{4}} < G < T(\log T)^{-1}$, where c_j, $1 \le j \le 4$, are positive absolute constants. This implies the resolution of the sixth power moment problem,

$$\int_0^T |\zeta(\tfrac{1}{2}+it)|^6 dt \ll T(\log T)^c,$$

which would have a tremendous impact on all of analytic number theory.

So how much can we hope for the proof of (5.4.5)? We have unfortunately no clues here. On the contrary one may even cast doubt on the conjecture (5.4.5) itself. There is in fact a possibility that a real ramification in the theory of $\mathscr{Z}_k(g)$ starts at $k = 3$. This can be seen by the following reasoning: If we try to extend the dissection argument given in (4.2.4) to $\mathscr{Z}_3(g)$ we encounter the sum

$$\sum_{m,n=1}^{\infty} \sigma(m;\alpha,\beta)\sigma(m+n;\gamma,\delta)W(m,n),$$

where W is a smooth weight and

$$\sigma(n;\alpha,\beta) = \sum_{d_1 d_2 | n} d_1^{\alpha} d_2^{\beta}.$$

What is remarkable here is that we have an analogue of Ramanujan's identity (1.1.14) for the arithmetic function $\sigma(n;\alpha,\beta)$. This was discovered by Bump [5] in the theory of the Fourier expansion of the minimal parabolic Eisenstein series for $SL(3,\mathbb{Z})$ (see also Vinogradov and Takhtadjan [72]). Hence $\mathscr{Z}_3(g)$ is definitely related to the theory of $SL(3,\mathbb{Z})$. We then invoke the fact that the algebra of the $SL(3,\mathbb{Z})$-invariant differential operators on the relevant homogeneous space has two generators, whereas in the case of $SL(2,\mathbb{Z})$ we have a unique generator that is the non-Euclidean Laplacian Δ. Hence the conjectural expression (5.4.5) might be replaced by a double sum over two kinds of independent objects. But it is hard to surmise the form of the double sum. Because of these it is highly desirable to have an honest extension to $SL(3,\mathbb{Z})$ of the theory developed in Chapters 1–3, even though it is not yet perfectly clear if such an extension has anything to do with $\mathscr{Z}_3(g)$.

At the same time it is also conceivable that the theory of $\mathscr{Z}_k(g)$ will eventually turn out to be more chaotic than what can be gathered from the above. This is due to the fact (Motohashi [49]) that the eighth power moment $I_4(T,G)$ admits an expression in terms of the objects pertaining to automorphic forms over the full modular group. Let us explain this phenomenon briefly: We try to correct the trivial inequality

$$I_k(T,G)^2 \le I_{2k}(T,G)$$

to an identity using a spectral theory. For this purpose let us put

$$\varphi_n(t) = (2^n n! \sqrt{\pi})^{-1/2} e^{-\frac{1}{2}t^2} h_n(t), \quad n = 0, 1, 2 \ldots,$$

where h_n is the nth Hermite polynomial. We then have the Fourier–Hermite expansion (see Lebedev [38, Chapter 4])

$$|\zeta(\tfrac{1}{2} + i(T+Gt))|^{2k} e^{-\frac{1}{2}t^2} = \sum_{n=0}^{\infty} z_n(T,G;k)\varphi_n(t),$$

where

$$z_n(T,G;k) = \int_{-\infty}^{\infty} \varphi_n(t) |\zeta(\tfrac{1}{2} + i(T+Gt))|^{2k} e^{-\frac{1}{2}t^2} dt.$$

The Parseval formula gives the identity

$$I_{2k}(T,G) = \sum_{n=0}^{\infty} z_n(T,G;k)^2. \tag{5.4.6}$$

We note that $z_0(T,G;k) = I_k(T,G)$. Thus (5.4.6) is a correction of the above inequality. We next invoke the expansion

$$e^{-(t-\xi)^2 + \frac{1}{2}t^2} = \pi^{\frac{1}{4}} \sum_{n=0}^{\infty} \frac{\varphi_n(t)}{\sqrt{n!}} \left(\frac{\xi}{\sqrt{2}}\right)^n,$$

which is equivalent to the generating series of h_n's. Hence we have

$$I_k(T+G\xi, G) = \pi^{-\frac{1}{4}} \sum_{n=0}^{\infty} \frac{z_n(T,G;k)}{\sqrt{n!}} \left(\frac{\xi}{\sqrt{2}}\right)^n.$$

The combination of this and (5.1.1) gives immediately a spectral expansion of each $z_n(T,G;2)$. Inserting the result into (5.4.6), $k=2$, we obtain the above assertion on the eighth power moment. It should, however, be added that the value of (5.4.6) is that of being theoretical. In practical use we need a certain truncation of the series on the right side, which is obviously a non-trivial problem. In this context it will perhaps be more worthwhile to consider the application of the Parseval formula for Mellin transforms to the function $Z_2(\omega)$.

We then focus our attention on the structural assertion stated in Theorem 5.4. We shall show that Theorem 5.4 provides us with a means to compose individual Hecke L-functions in terms of the zeta-values, though on a generally believed hypothesis. Hence the zeta-function not only is served by those L-functions but also serves them in return:

We suppose that the eigenvalue $\lambda_j = \kappa_j^2 + \frac{1}{4}$ is simple (cf. Sarnak [64]). Then we have, by (5.3.13),

$$\lim_{\omega \to \frac{1}{2}+i\kappa_j} (\omega - \tfrac{1}{2} - i\kappa_j) Z_2(\omega)$$

$$= (\tfrac{1}{2}\pi)^{\frac{1}{2}} \left(2^{-i\kappa_j} \frac{\Gamma(\frac{1}{2}(\frac{1}{2} - i\kappa_j))}{\Gamma(\frac{1}{2}(\frac{1}{2} + i\kappa_j))} \right)^3 \Gamma(2i\kappa_j) \cosh(\pi\kappa_j) \alpha_j H_j(\tfrac{1}{2})^3. \tag{5.4.7}$$

This may be understood to imply that $\frac{1}{2} + i\kappa_j$ can be found among the poles of $Z_2(\omega)$, and that from the corresponding residue we may fix the value of $\alpha_j H_j(\frac{1}{2})^3$. Thus it is possible to evaluate $H_j(\frac{1}{2})$ by means of the zeta-function, provided we can find the first Fourier coefficient of the Maass wave ψ_j. Actually we are able to extend this fact to $H_j(s)$, $\frac{1}{2} \le \operatorname{Re} s < 1$, by considering, instead of $Z_2(\omega)$, the expression

$$Z_2(\omega, s) = \int_1^\infty \left[\zeta\left(\tfrac{1}{2}(s + \tfrac{1}{2}) + it\right) \zeta\left(\tfrac{1}{2}(s + \tfrac{1}{2}) - it\right) \right]^2 t^{-\omega} dt.$$

Following closely the argument of Sections 4.7 and 5.3 one can find that $Z_2(\omega, s)$ has a simple pole at $\omega = 1 - s + i\kappa_j$. The residue is equal to the right side of (5.4.7) but with the factor $H_j(\frac{1}{2})^3$ being replaced by $H_j(\frac{1}{2})^2 H_j(s)$.

Thus, in principle, the functional properties of $H_j(s)$ in the critical strip can be expressible in terms of those of the zeta-function, provided λ_j is simple and $H_j(\frac{1}{2}) \neq 0$. This implies further that the corresponding cusp-form ψ_j could be composed from the zeta-values. For, according to Epstein, Hafner and Sarnak [10], we have the representation

$$\psi_j(re^{i\theta}) = \frac{1}{2\pi} \rho_j(1) \int_{-\infty}^{\infty} H_j^*(\tfrac{1}{2} + it) r^{it} k(\theta, t; \kappa_j) dt \quad (r > 0, -\pi < \theta < \pi). \tag{5.4.8}$$

Here $\rho_j(1)$ is the first Maass–Fourier coefficient of ψ_j,

$$H_j^*(s) = \frac{1}{4} \pi^{-s} \Gamma(\tfrac{1}{2}(s + i\kappa_j)) \Gamma(\tfrac{1}{2}(s - i\kappa_j)) H_j(s),$$

and

$$k(\theta, t; u) = (\sin\theta)^{-it} F\left(\tfrac{1}{2}(\tfrac{1}{2} + i(t+u)), \tfrac{1}{2}(\tfrac{1}{2} + i(t-u)); \tfrac{1}{2}; -(\cot\theta)^2\right)$$

with the hypergeometric function F.

Now the following diagram of interactions becomes evident:

$$\begin{array}{ccc} \zeta(s) & \longleftrightarrow & \{H_j(s)\} \\ \updownarrow & & \updownarrow \\ E(z,s) & \overset{?}{\leftarrow \cdots \rightarrow} & \{\psi_j(z)\} \end{array} \tag{5.4.9}$$

Here it would be better to replace $\zeta(s)$ by $\zeta^2(s)$ for the sake of accuracy. Also the set of cusp-forms should better be taken for that of even forms (note that $E(z,s)$ is even). The left and right columns do not require explanation; and the top row is what we have observed above. As is indicated in the bottom row, we think it is likely that there exists a direct linkage between the Eisenstein series and even cusp-forms. The actual situation will probably be clarified with a suitable extension of the arguments given in Section 2.6. The right column can of course be decomposed according to the spectrum of Δ. At each eigen-space the Hecke correspondence takes place. Hence the explicit formula (4.7.1) or rather the spectral decomposition (4.7.3) is a means to show those infinitely many correspondences at once. The diagram (5.4.9) is, however, incomplete. What is missing there is the rôle of holomorphic cusp-forms. Although their contribution is nominal, we should fix what they actually stand for in the theory of mean values of the zeta-function.

We now turn to another aspect of the mean-value problem. We wonder if the two omega results (5.3.15) and (5.4.4) have any extensions to higher power moments. To make the situation explicit we define $E_k(T)$ analogously to the above $E_1(T)$ and $E_2(T)$ so that

$$\int_0^T |\zeta(\tfrac{1}{2}+it)|^{2k}dt = TP_k(\log T) + E_k(T).$$

Here P_k is a certain polynomial depending only on k. It should be stressed that we do not take this for an asymptotic formula. In fact, only the cases $k = 1, 2$ have so far been established as such; and it is even possible that this formulation does not have much sense, especially for $k \geq 4$ as we are going to indicate.

We think it is highly plausible to have

$$E_3(T) = \Omega(T^{\frac{3}{4}-\delta}) \tag{5.4.10}$$

for any fixed $\delta > 0$, which is a simple extrapolation of (5.3.15) and (5.4.4). To illustrate this conjecture we note that it is known (see Ivić [19, p. 172]) that for any $k \geq 1$

$$\sup_{|t|\leq T} |\zeta(\tfrac{1}{2}+it)| \ll \Big(\sup_{V\leq 2T} |E_k(V)| \log T \Big)^{1/(2k)}.$$

Thus, if we suppose the contrary of (5.4.10) for some δ, we would have

$$\zeta(\tfrac{1}{2}+it) \ll (|t|+1)^{\frac{1}{8}-\eta}$$

for an $\eta < \frac{1}{6}\delta$. This seems, however, too good for what we can expect from the theory of the mean values of $\zeta(s)$. In fact, the combination of (5.2.21), (5.3.15), (5.4.3), and (5.4.4) strongly suggests to us that the best estimate we can extract from the cases $k = 1, 2$ would be

$$\zeta(\tfrac{1}{2}+it) \ll (|t|+1)^{\frac{1}{8}+\varepsilon}$$

for any $\varepsilon > 0$. This itself is fantastic in view of the enormous difficulties that have been experienced in bounding $\zeta(s)$. Thus it is unrealistic to expect the sixth power moment problem will be settled in a way that denies (5.4.10). If, on the contrary, it ever turns out that the conjecture (5.4.10) is false, we would certainly welcome the event, for only a truly extraordinary method would be able to achieve it.

On the other hand, what will the situation for $k \geq 4$ be? The case $k = 5$ inevitably touches the Riemann Hypothesis, though in an awkward way: If the linearity among the exponents in (5.3.15), (5.4.4), and (5.4.10) continues to hold up to $k = 5$, then the Lindelöf hypothesis and consequentially the Riemann hypothesis would be denied. This is hard to believe. But then, how will this linearity cease to hold? This is also hard to speculate on. Some people set out conjectures on this issue. We think, however, it is premature to discuss them. What we wanted to indicate in the above is that (5.3.15) seems to stand for a threshold, beyond which the zeta-function will start unveiling itself. To be well prepared for the event we shall certainly need a broader stage and finer tuning of instruments.

References

[1] A.O. Atkin and J. Lehner. Hecke operators on $\Gamma_0(m)$. *Math. Ann.*, **185** (1970), 134–160.

[2] F.V. Atkinson. The mean value of the Riemann zeta-function. *Acta Math.*, **81** (1949), 353–376.

[3] E. Bombieri. *Le Grand Crible dans la Théorie Analytique des Nombres*. Astérisque vol. **18**, Soc. Math. France, Paris, 1987/1974.

[4] R.W. Bruggeman. Fourier coefficients of cusp forms. *Invent. math.*, **45** (1978), 1–18.

[5] D. Bump. *Automorphic Forms on* $\mathrm{SL}(3,\mathbb{R})$. Lect. Notes in Math., vol. **1083**, Springer-Verlag, Berlin, 1984.

[6] D. Bump et al. An estimate for Fourier coefficients of Maass wave forms. *Duke Math. J.*, **66** (1992), 75–81.

[7] R. Courant and D. Hilbert. *Methods of Mathematical Physics*. Vol. I. Interscience Publ. Inc., New York, 1953.

[8] J.-M. Deshouillers and H. Iwaniec. Kloosterman sums and Fourier coefficients of cusp forms. *Invent. math.*, **70** (1982), 219–288.

[9] —. Power mean-values for Dirichlet's polynomials and the Riemann zeta-function. I, II. *Mathematika*, **29** (1982), 202–212; *Acta Arith.*, **43** (1984), 305–312.

[10] C. Epstein et al. Zeros of L-functions attached to Maass forms. *Math. Z.*, **190** (1985), 113–128.

[11] J.D. Fay. Fourier coefficients of the resolvent for a Fuchsian group. *J. Reine Angew. Math.*, **293–294** (1973), 143–203.

[12] D. Goldfeld and P. Sarnak. Sums of Kloosterman sums. *Invent. math.*, **71** (1983), 243–250.

[13] A. Good. Beitraege zur Theorie der Dirichletreihen die Spitzenformen zugeordenet sind. *J. Number Theory*, **13** (1981), 18–65.

[14] —. The square mean of Dirichlet series associated with cusp forms. *Mathematika*, **29** (1982), 278–295.

[15] I.S. Gradshteyn and I.M. Ryzhik. *Table of Integrals, Series and Products*. Academic Press, London, 1980.

[16] G.H. Hardy. *Ramanujan*. Cambridge University Press, London, 1940.

[17] E. Hecke. *Mathematische Werke*. Vandenhoeck & Ruprecht, Göttingen, 1970.

[18] D.A. Hejhal. *The Selberg Trace Formula for* $\mathrm{PSL}(2,\mathbb{R})$. II. Lect. Notes in Math., vol. **1001**, Springer-Verlag, Berlin, 1983.

[19] A. Ivić. *The Riemann Zeta-Function*. John Wiley & Sons, Inc., New York, 1985.

[20] —. *Lectures on the Mean Values of the Riemann Zeta-Function*. Tata Inst. Fund. Res. Lectures on Math. and Phys., vol. **82**, Springer-Verlag, Berlin, 1991.

[21] A. Ivić and Y. Motohashi. The mean square of the error term in the fourth moment of the zeta-function. *Proc. London Math. Soc.*, (3) **69** (1994), 309–329.

[22] H. Iwaniec. Fourier coefficients of cusp forms and the Riemann zeta-function. Exp. No. 18, Sém. Th. Nombres, Univ. Bordeaux 1979/80.

[23] —. On mean values for Dirichlet's polynomials and the Riemann zeta-function. *J. London Math. Soc.*, (2) **22** (1980), 39–45.

[24] —. Non-holomorphic modular forms and their applications. In *Modular Forms* (Symp., Durham, England, 1983; R.A. Rankin, editor), Ellis Horwood, Chichester, 1984, pp. 157–196.

[25] —. Promenade along modular forms and analytic number theory. In *Topics in Analytic Number Theory* (S. W. Graham and J. D. Vaaler, editors), Univ. Texas Press, Austin, 1985, pp. 221–303.

[26] —. The spectral growth of automorphic L-functions. *J. Reine Angew. Math.*, **428** (1992), 139–159.

[27] —. *Introduction to the Spectral Theory of Automorphic Forms*. Bibl. de la Revista Mat. Iberoamer., Madrid, 1995.

[28] M. Jutila. The fourth moment of Riemann's zeta-function and the additive divisor problem. In *Analytic Number Theory, Proceedings of a Conference in Honor of Heini Halberstam* (B.C. Bernt, H.G. Diamond, A.J. Hildebrand, editors), Birkhäuser, Boston–Basel–Berlin, 1996, pp. 515–536.

[29] —. The additive divisor problem and its analogues for Fourier coefficients of cusp forms. I, II. To appear in *Math. Z.*

[30] —. Atkinson's formula revisited. Preprint

[31] —. On spectral large sieve inequalities. Preprint.

[32] —. Mean values of Dirichlet series via Laplace transforms. Preprint.

[33] M. Jutila and Y. Motohashi. Mean value estimates for exponential sums and L-functions: a spectral theoretic approach. *J. Reine Angew. Math.*, **459** (1995), 61–87.

[34] S. Katok and P. Sarnak. Heegner points, cycles and Maass forms. *Israel J. Math.*, **84** (1993), 193–227.

[35] N.V. Kuznetsov. Petersson hypothesis for forms of weight zero and Linnik hypothesis. Preprint, Khabarovsk Complex Res. Inst., East Siberian Branch Acad. Sci. USSR, Khabarovsk, 1977. (Russian)

[36] —. Petersson hypothesis for parabolic forms of weight zero and Linnik hypothesis. Sums of Kloosterman sums. *Math. USSR–Sb.*, **39** (1981), 299–342.

[37] —. Convolution of the Fourier coefficients of the Eisenstein-Maass series. *Zap. Nauchn. Sem. LOMI*, **129** (1983), 43–84. (Russian)

[38] N.N. Lebedev. *Special Functions and their Applications*. Dover Publications, New York, 1972.

[39] Ju.V. Linnik. The large sieve. *Dokl. Akad. Nauk SSSR*, **30** (1941), 292–294. (Russian)

[40] —. Additive problems and eigenvalues of the modular operators. *Proc. Internat. Congress Math. Stockholm*, 1962, pp. 270–284.

[41] H. Maass. Über eine neue Art von nichtanalytischen automorphen

Funktionen und die Bestimmung Dirichletscher Reihen durch Funktionalgleichungen. *Math. Ann.*, **121** (1949), 141–183.

[42] T. Meurman. On the order of the Maass L-function on the critical line. In *Coll. math. Soc. J. Bolyai 51. Number Theory, Budapest 1987*, North-Holland, Amsterdam, 1989, pp. 325–354.

[43] J. Milnor. Hyperbolic geometry: the first 150 years. *Bull. Amer. Math. Soc.*, **6** (1982), 9–24.

[44] Y. Motohashi. *Sieve Methods and Prime Number Theory*. Tata Inst. Fund. Res. Lectures on Math. and Phys., vol. **72**, Springer-Verlag, Berlin, 1983.

[45] —. *Riemann–Siegel Formula*. Ulam Chair Lectures, Colorado Univ., 1987.

[46] —. Spectral mean values of Maass wave form L-functions. *J. Number Theory*, **42** (1992), 258–284.

[47] —. An explicit formula for the fourth power mean of the Riemann zeta-function. *Acta Math.*, **170** (1993), 181–220.

[48] —. The binary additive divisor problem. *Ann. Sci. l'Ecole Norm. Sup. 4e série*, **27** (1994), 529–572.

[49] —. A note on the mean value of the zeta and L-functions. VIII. *Proc. Japan Acad.*, **70**A (1994), 190–193.

[50] —. The mean square of Hecke L-series attached to holomorphic cusp-forms. *RIMS Kyoto Univ. Kokyuroku*, **886** (1994), 214–227.

[51] —. On Kloosterman-sum zeta-function. *Proc. Japan Acad.*, **71**A (1995), 69–71.

[52] —. The Riemann zeta-function and the non-Euclidean Laplacian. *Amer. Math. Soc. Sugaku Expositions*, **8** (1995), 59–87.

[53] —. A relation between the Riemann zeta-function and the hyperbolic Laplacian. *Ann. Scuola Norm. Sup. Pisa Cl. Sci.*, (4) **22** (1995), 299–313.

[54] —. On Kuznetsov's trace formulas. In *Analytic Number Theory, Proceedings of a Conference in Honor of Heini Halberstam* (B.C. Bernt, H.G. Diamond, A.J. Hildebrand, editors) Birkhäuser, Boston–Basel–Berlin, 1996, pp. 641–667.

[55] —. The Riemann zeta-function and the Hecke congruence subgroups. *RIMS Kyoto Univ. Kokyuroku*, **958** (1996), 166–177.

[56] —. The mean square of Dedekind zeta-functions of quadratic number fields. In *Sieve Methods, Exponential Sums and their Applications in Number Theory* (G.R.H. Greaves, G. Harman, M.N. Huxley, editors) Cambridge University Press, Cambridge, 1997, pp. 309–324.

[57] —. Trace formula over the hyperbolic upper half-space. Preprint.

[58] D. Niebur. A class of nonanalytic automorphic functions. *Nagoya Math. J.*, **52** (1973), 133–145.

[59] H. Petersson. Über eine Metrisierung der ganzen Modulformen. *Jber. Deutsch. Math. Verein.*, **49** (1939), 49–75.

[60] N.V. Proskurin. Summation formulas for general Kloosterman sums. *Zap. Nauchn. Sem. LOMI*, **82** (1979), 103–135. (Russian)

[61] R.A. Rankin. *Modular Forms and Functions*. Cambridge University Press, Cambridge, 1977.

[62] F. Riesz and B. Sz.-Nagy. *Functional Analysis*. Frederick Ungar Publ. Co., New York, 1955.

[63] W. Roelcke. Über die Wellengleichung der Grenzkreisgruppen erster Art. *Sitz. Ber. Heidelberger Akad. Wiss. Math. Nat. Kl.*, 4 Abh., 1956.

[64] P. Sarnak. *Arithmetic Quantum Chaos*. R.A. Blyth Lectures, University of Toronto, 1993.

[65] A. Selberg. Harmonic analysis and discontinuous groups in weakly symmetric Riemannian spaces with applications to Dirichlet series. *J. Indian Math. Soc.*, **20** (1956), 47–87.
[66] —. On the estimation of Fourier coefficients of modular forms. *Proc. Symp. Pure Math.*, **8** (1965), 1–15.
[67] C.L. Siegel. *Topics in Complex Function Theory*. Vol. II. Wiley-Interscience, New York, 1971.
[68] N. Sonine. Recherches sur les fonctions cylindriques et le développement des fonctions continues en séries. *Math. Ann.*, **16** (1880), 1–80.
[69] E.C. Titchmarsh. *Introduction to the Theory of Fourier Integrals*. Clarendon Press, Oxford, 1948.
[70] —. *The Theory of the Riemann Zeta-Function*. Clarendon Press, Oxford, 1951.
[71] N.J. Vilenkin. *Special Functions and the Theory of Group Representations*. Amer. Math. Soc. Transl. Math. Monographs, vol. **22**, Providence, R.I., 1968.
[72] A.I. Vinogradov and L.A. Takhtadjan. Theory of Eisenstein series for the group $SL(3,\mathbb{R})$ and its application to a binary problem. *J. Soviet Math.*, **18** (1982), 293–324.
[73] —. The zeta-function of the additive divisor problem and the spectral decomposition of the automorphic Laplacian. *Zap. Nauchn. Sem. LOMI*, **134** (1984), 84-116. (Russian)
[74] G.N. Watson. *A Treatise on the Theory of Bessel Functions*. Cambridge Univ. Press, London, 1944.
[75] E.T. Whittaker and G.N. Watson. *A Course of Modern Analysis*. Cambridge Univ. Press, London, 1927.
[76] E. Yoshida. Remark on the Kuznetsov trace formula. Preprint.

Author index

Subject index